D0041315

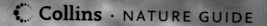
Collins · NATURE GUIDE

WILD ANIMALS
OF BRITAIN & EUROPE
Helga Hofmann

Translated by
MARTIN WALTERS
Scientific Consultant Dr Gordon Corbet

Collins

Reprinted in 2010 for
Independent Book Sales

HarperCollins Publishers Ltd.
77-85 Fulham Palace Road
London W6 8JB

Collins is a registered trademark of
HarperCollins Publishers Ltd.

11 10
11

Written by Helga Hofmann
Drawings by Heinz Bogner

This edition translated by Martin Walters
Scientific consultant: Dr Gordon Corbet

The translator would like to thank Teresa Sheppard for her help
with typing the translation.

Originally published in German as a GU Nature Guide
by Gräfe und Unzer GmbH, Munich

© Gräfe und Unzer GmbH, Munich, 1988
© in this English translation HarperCollins Publishers, 1995

ISBN-13: 978 0 00 762727 1

Collins uses papers that are natural, renewable and recyclable
products made from wood grown in sustainable forests. The
manufacturing processes conform to the environmental
regulations of the country of origin.

Printed and bound in China by South China printing company

Colour	Symbol	Mammal group	Page number
		Insectivores: hedgehogs, shrews, moles	8–25
		Bats: horseshoe bats, vesper bats, free-tailed bats	26–47
		Primates: Barbary Ape	48–49
		Rodents: squirrels, hamsters, lemmings, voles, beavers, Coypu, dormice, mice, birch mice, Porcupine	50–111
		Lagomorphs: hares and rabbits	112–119
		Carnivores: weasel and relatives, Raccoon, bears, Genet, mongooses, dogs, cats	120–173
		Pinnipeds: seals and Walrus	174–185
		Hoofed mammals (odd-toed): Horse	186–187
		Hoofed mammals (even-toed): pigs, deer, bovids (cattle, Chamois, Ibex, goats, sheep, Musk Ox)	188–227
		Whales: toothed whales (dolphins, white whales, sperm whales), baleen whales (rorquals, right whales)	228–247

Introduction

This is a book for everyone interested in the wild animals of Britain and Europe. The handy format combined with the colour-coded key make it an ideal reference for those who wish to identify and learn about mammals. Almost all the wild mammals of Europe are illustrated and described.

There are 225 colour photographs that capture the animals in their natural settings, showing distinctive features or typical behaviour. In addition, 180 drawings show details of body shape or highlight important differences between similar species, special behaviour patterns or other signs, such as tracks, nests or burrows.

A novel identification system has been devised for this book. It is effective and simple to use. The mammals are divided into six groups, each with its own particular colour. Coloured thumb-markers at the edge of the right-hand pages simplify finding the various animal types. Symbols within these markers are a further aid in designating the animal group – they show the type of animal found on those pages (see key on p. 5). Once the correct group has been located, the clear colour photographs and helpful drawings make it easy to identify the animals and to distinguish between similar species.

This book is more than merely a guide to identification however. It also contains a wealth of fascinating detail about the lives of our wild European mammal species – their behaviour, their physiology, the habitats in which they live, and their many relations with people.

Species included

Almost all native and wild European mammals are included, excluding domesticated species.

A number of mammal species have been introduced to Europe by people in the past – either deliberately, to increase the range of animals for hunting, or inadvertently, the animals having escaped from captivity. Some of these found the conditions suited them so well that they built up stable or in some cases growing populations, Muskrat and Raccoon for example. Such species are included here as an established part of the European mammal fauna. Exceptions are Wild Horse and Aurochs, both extinct as true wild species, but included because of their importance as ancestors of horses and cattle.

Whale species that turn up regularly (every year, or at certain seasons) in European coastal waters have also been included.

The descriptions

Even though the photographs have, almost without exception, been taken in the wild, they cannot always show all the characters of a species as clearly as a drawing. The photographs illustrate the typical postures or behaviour of the particular species. Mammals do however show individual variation, even within the same species, and for this reason it is wise not to rely solely on a photograph when making an identification, but use a combination of the photograph, any additional drawing and the diagnostic features mentioned in the text.

Each description is divided into two sections: the notes and the main text. If more than one English name is commonly used for the species, they are mentioned. The notes contain all the important details for identification, under the following headings:

Identification
Normal average body size and weight of the adult (not extremes). These are divided into male and female where there is a significant difference between the sexes. Description of typical appearance, with a mention of any particularly noteworthy features.

Abbreviations used:

HB: head–body length (not including tail)
T: tail length
W: body weight
WS: wingspan (bats)
S: shoulder height
L: total length (whales)

Distribution
For most this can only be a rough indication, since each species only occurs in particular habitats within its overall geographic range. If a species also occurs outside Europe this is noted.

Habitat
That most typical for each species.

Behaviour
Time of main activity – whether day or night. Whether it lives in groups or tends to be solitary. What kind of nest it builds or home it has and how it spends the winter.

Food
The commonest types and sources of food.

Breeding
The normal time of year for mating is indicated, as well as the gestation period, number of births, average number of young per birth and length of time of development for young, until they are self-sufficient.

The main text has interesting information about the species, as well as details about the biology and characteristics of each group. The descriptions should help to further our understanding of mammals and thereby increase our sense of responsibility towards them. Human interest is essential for their conservation and protection; many species are already threatened in the wild.

Mammal Biology
The short section on mammal biology towards the end of the book (p. 248) sets out the scientific background of mammals in brief, as well as explaining the technical terms used in the main text. This section also includes information on some mammal conservation organisations.

Western Hedgehog

(Hedgehog Family) *Erinaceus europaeus*

Identification: HB 22-30 cm, T 2-5 cm, W 500-1900 g. Covered with brown and white striped prickles, head and underside with rough, yellow-brown hairs; head pointed, with small eyes and rounded, leathery ears.

Distribution: Western Europe including southern Scandinavia and Finland; widespread in Britain.

Habitat: Bushy woodland margins, parks, gardens; to 2000 m in mountains.

Behaviour: Most active at dusk and by night, although the young are sometimes active during the daytime; solitary; climbs and digs well but swims for only short distances; rolls up into a ball when threatened; by day hides in thick vegetation or underneath stones; nest lined with grass and leaves; hibernates Oct-Apr.

Food: Insects, worms, snails and slugs; also young mice, frogs, snakes and occasionally fruits and berries.

Breeding: Mating season Apr-Aug; gestation 5-6 weeks; 1-2 litters per year, each with 4-7 blind and helpless young; newly born young have white, soft prickles and open their eyes and ears in about 2 weeks; suckled for 3-4 weeks and independent when about 6 weeks old; male not involved in rearing the young.

The prickles which make the Hedgehog so unmistakable are modified back hairs. A fully grown Hedgehog has about 16,000 of them. Normally these lie along the body facing backwards but they can be raised using powerful layers of muscles. The muscles cover the entire back of the Hedgehog rather like the shell of a tortoise. When the Hedgehog is alarmed, and also during hibernation, these muscles contract and the spines are raised. At the same time the head and legs are pulled safely in and tucked underneath so that the whole animal becomes a spiny ball on which it is very difficult for any attacker to get a secure grip. However, this alarm reaction is no defence against traffic.

The Hedgehog uses its muscular covering when rolling up into a ball (shown here without the skin on the back)

Hedgehogs lose a lot of warmth through their spines and for this reason they are often attracted (particularly during the cool of the evening) to roads whose asphalt holds the heat of the day. Hundreds of thousands are killed every year on our roads. Another danger to Hedgehogs are chemical sprays used on field crops and also in gardens against weeds and insect pests. These often accumulate in garden invertebrates which are in turn eaten by Hedgehogs who may themselves then suffer poisoning. This is particularly unfortunate because Hedgehogs do a lot of good in the garden by eating pests. (Continued p. 10.)

Female Western Hedgehog with young (above)
Half-grown young Western Hedgehogs (below)

Western Hedgehog (continued)

Hedgehogs tend to be attacked by many parasites such as ticks, mites and roundworms. Hedgehog fleas which are often numerous in amongst the spines are particular to Hedgehogs and do not transfer to household pets or to people.

Rival males tend to be aggressive towards each other. They raise their neck spines and each tries to get under the other to inflict a bite to the unprotected belly. Such fights are accompanied by much hissing and wheezing, but the participants are rarely badly injured.

Hedgehogs pair soon after the end of hibernation. Like most mammals, Hedgehogs mate from behind and not belly to belly as was previously believed. During mating the female Hedgehog keeps her spines as flat as possible. The male is firmly expelled from the vicinity of the nest long before the young are born and the mother alone provides for the young.

During the birth, the tiny soft prickles of the young are protected by their skin and therefore do not damage the mother. In the first few days the skin unfolds to reveal the prickles which remain soft for a few more days. The young begin to explore with their mother when they are about three weeks old, and by six weeks they are self-sufficient. At this stage the first long adult spines appear. Hedgehogs sometimes have a second litter in late summer but these often do not survive the winter. In order to overwinter successfully they need to be about 500-700 g in weight. Very young Hedgehogs are occasionally found late in the year and many people try to see them through the winter by putting them indoors, but often without success. Hedgehogs do not store food for the winter but exist solely on their fat reserves. When they wake up again the following spring they have lost about a quarter of their body weight.

The **Eastern Hedgehog** (*Erinaceus concolor*), distinguished by having white on the breast and throat, replaces the Western Hedgehog in eastern Europe and extends as far as Turkey and Israel.

The North African or **Algerian Hedgehog** (*Erinaceus algirus*) also occurs in Spain, Balearic Islands, and in southern France. It is noticeably paler than the European Hedgehog and has a wider parting in the spines on its forehead.

Hedgehog stretching – they very often do this on waking

Hedgehogs can swim when really necessary (above)
Alert Hedgehog sniffing the air (below)

Pygmy Shrew

Sorex minutus

(Shrew Family)

Identification: HB 4.5-6.5 cm, T 3-4.5 cm, W 3-7 g. Resembles Common Shrew (p. 14) but smaller; snout narrower; tail relatively longer.

Distribution: The whole of Europe except the western Mediterranean and Iceland; temperate parts of Asia to China. Widespread in Britain and Ireland.

Habitat: Woodland edges, hedgerows, damp meadows, gardens, parks; in winter sometimes also in buildings.

Behaviour: Active by night and day; solitary; hunts for food mostly above ground, more rarely in mouse burrows; does not burrow for itself.

Food: Beetles and other insects, spiders, worms, slugs and snails.

Breeding: Mating season Apr-Aug; gestation 19 days; 1-2 litters per year, each with 4-8 young. Development as for Common Shrew.

Shrews are often mistaken for mice, although they are not closely related. Shrews belong to the insectivores, a much older group than the mice in evolutionary terms.

The Pygmy Shrew is one of our smallest mammals. Such very small animals have a high surface area for their overall size and therefore lose a lot of warmth from their bodies. In order to counterbalance this temperature loss they have to take in a lot of energy in the form of food. The Pygmy Shrew begins to starve after only nine hours without food. Every two hours or so they hunt feverishly amongst taller grasses and ground vegetation, taking very short naps between feeding bouts. Like all shrews they do not hibernate, but continue to feed throughout the winter, often underneath the snow.

Alpine Shrew

Sorex alpinus

(Shrew Family)

Identification: HB 6-7.5 cm, T 6-7.5 cm, W 6-10 g. Coat colour slate grey, slightly paler beneath, with whitish feet; snout long and pointed; tail about the same length as the body.

Distribution: Mountains of central Europe, Alps, Pyrenees, Carpathians, Balkans, 500-2500 m.

Habitat: Coniferous woods; damp meadows; mountain streams.

Behaviour: Active by night and day; solitary.

Food: Insects, spiders, woodlice, slugs, snails and worms.

Breeding: Mating season Apr-Aug; breeding biology similar to Common Shrew (p. 14).

This species belongs, with the Pygmy and Common Shrews, to the red-toothed shrew group – the tips of the teeth are dark red. The jaws of shrews are quite different from those of mice. Whereas the rodents have characteristic long, chisel-like cutting teeth at the front of the jaw, a shrew's jaws are more like those of a miniature carnivore, with rows of sharp teeth.

Skull, showing teeth: red-toothed shrew (left), white-toothed shrew (middle), and mouse (right).

Pygmy Shrew (above)
Alpine Shrew (below)

Common Shrew

Sorex araneus

(Shrew Family)

Identification: HB 6-8.5 cm, T 3-5 cm, B 6-12 g. Dark brown above, with chestnut flanks and yellow or silvery grey underneath. Snout narrow and sharply pointed; ears round, just projecting above coat; tail with short hairs.

Distribution: Whole of Europe except the Mediterranean lowlands, Ireland and Iceland; temperate parts of Asia. Widespread in Britain.

Habitat: Damp marshy places, woodland, gardens, parks; in winter occasionally in buildings.

Behaviour: Active by day and night, solitary. Seeks food in leaf litter and humus, also sometimes climbs up into shrubs. Swims well. It either digs out its own nest, or uses abandoned mouse burrows.

Food: Insects, spiders, slugs, snails, worms, young mice and carrion.

Breeding: Mating season Apr-Nov; gestation 19-21 days; 2-4 litters per year, each with 5-9 blind, naked young. These first open their eyes at around 18 days, and are suckled for 3 weeks before they are independent.

Our commonest and best-known shrew. Very active like all shrews, moving hectically to and fro through thick ground vegetation in its frantic search for food. The tiny eyes make it very short-sighted, but it has an acute sense of smell and also keen hearing, even though the ears are almost covered by fur. Takes a wide variety of food, including beetles and other insects, but also dead vertebrates, including members of the same species (during times of food shortage it will even eat its own offspring). The size of its prey varies between 1-2 mm, in the case of the smallest insects to up to 10 cm long earthworms – much longer than the shrew itself.

A shrew making a meal of an earthworm

The young shrews are born in a nest lined with moss and leaves. At birth they weigh only about 1 g and their naked skin is so thin that the internal organs are clearly visible through it.

Shrews are mainly preyed upon by birds of prey and owls. Carnivorous mammals often kill shrews but then leave them uneaten. This is because shrews have glands on their flanks and at the base of the tail that produce a scent which is apparently unpleasant tasting for foxes, martens and other carnivorous mammals. Domestic cats also tend to leave shrews uneaten after they have killed them.

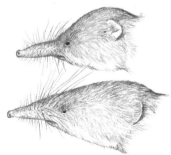

Comparison of heads of Pygmy Shrew (above) and Common Shrew (below)

The ears of a Common Shrew are scarcely visible (above) Shrews are perpetually hungry (below)

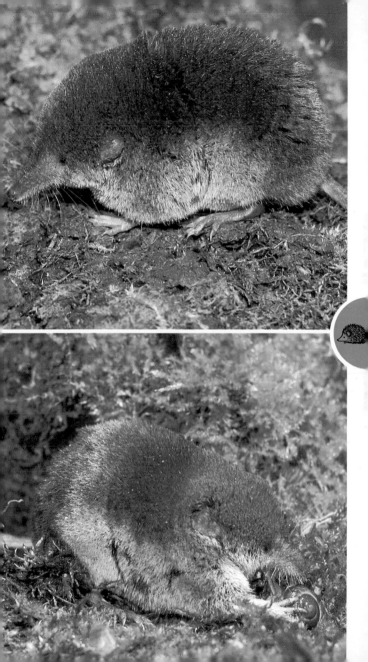

Water Shrew

Neomys fodiens

(Shrew Family)

Identification: HB 7-9.5 cm, T 5-7 cm, W 10-23 g. Coat silky, dark grey to black above and usually with a sharp border between this and the silver grey or white underside (underside rarely also black); ears just poking out of the coat; tail long, with a keel of white bristles along the underside. Hind feet large, with wide band of fringing hairs.

Distribution: Europe, except Ireland, Iceland and parts of the Mediterranean region; temperate parts of Asia; in mountains up to 2500 m. Widespread in Britain but rather scattered in the north.

Habitat: Clear, flowing or standing water with well-vegetated banks.

Behaviour: Active day and night, usually solitary, swims and dives exceptionally well and can also walk along the ground when submerged. Hunts in water and on land. Digs tunnels in the bankside vegetation and lines its nest with a thick layer of plant material.

Food: Aquatic insects, worms, slugs, snails, crustaceans, frogs, newts and small fish.

Breeding: Mating season Apr-Sep; gestation period 24-27 days; 2-3 litters per year, each with 4-8 naked blind young. Young open their eyes after about 3 weeks; are suckled for 4-5 weeks and are independent at 6 weeks.

Note: The bite of the Water Shrew is poisonous for small animals and leads to paralysis.

Water Shrews are the largest of our native shrews. Even though their feet are not webbed like those of many aquatic mammals, they are excellent swimmers. The hind feet with their stiff fringing hairs are like small paddles and the hairy tail also makes a fine rudder. When Water Shrews are submerged the hair traps a layer of air which insulates it from the cold of the water, giving the diving animal a silvery look. The air also increases the shrew's buoyancy so that it has to paddle rapidly to stay submerged. It uses its thin snout to probe underneath stones at the bottom of the water and to catch small invertebrates. However, Water Shrews can also tackle prey up to 60 times heavier than themselves. The poison in their bite is sufficiently strong to immobilise frogs or smaller fish. Food is carried to a hiding place on land to be eaten and is also often stored. The entrance tunnels are so narrow that most of the water is squeezed out of the shrew's fur as it returns to its lair.

Cross-section of Water Shrew nest in bankside vegetation. At least one entrance hole emerges under water

Miller's Water Shrew (*Neomys anomalus*) is found in mountainous parts of central and southern Europe. It is very similar to the Water Shrew but can also be found in areas without open water. In this species the swimming hairs on the tail are less well developed.

Water Shrew
Note fringe of hair on hind toes to aid swimming (below)

Greater White-toothed Shrew

(Shrew Family)   *Crocidura russula*

Identification: HB 6.5-9.5 cm, T 3.5-5 cm, W 6-14 g. Upperside grey to
chestnut brown, grading into a somewhat paler belly; feet whitish; ears
large; tail with long bristles.

Distribution: Central and southwest Europe; North Africa. In the British Isles
only found on certain Channel Islands (Alderney, Guernsey and Herm).

Habitat: Dry, sunny places, woodland edges, meadows, gardens and parks;
often found in houses, at any time of the year.

Behaviour: Active by day and night and usually solitary. Does not climb.
It either digs its own burrows or uses old mouse or mole tunnels. Nest
made of grass and leaves; in gardens often under compost heaps.
Female sometimes leads young in a caravan formation if the nest is
disturbed.

Food: Insects and their larvae, spiders, woodlice, slugs, snails, worms, mice
and dead vertebrates.

Breeding: Mating season Mar-Sep; gestation 31 days; 2-4 litters a year, each
with 3-9 young; development as for Bicoloured White-toothed Shrew
(p. 22).

This species, and the other white-toothed shrews, has completely white
teeth. The snout has short fur from which the long whiskers protrude. This
species is often found in gardens, sheds and cellars. Like other shrews it

needs to consume almost its
body weight of prey each day
and hunts for spiders, woodlice,
cockroaches and flies. It also
sometimes catches House Mice.
In gardens the Greater
White-toothed Shrew's main
food is slugs , snails, worms and
many insect larvae. In this way it
destroys a lot of garden pests.

Greater White-toothed Shrews eat large numbers
of insect pests and other invertebrates

Shrews, like rabbits and hares, have an unusual method of double
digestion called refection. The products of the large intestine are eaten
again and redigested. This special kind of faeces contains important
vitamins which are produced by the bacteria of the intestine.

When disturbed, the shrew will produce a sharp twittering call.
Territorial disputes involve quite a lot of vocalisation, noisy confrontations
and bared teeth.

Refection

The Greater White-toothed Shrew, like all shrews,
is short-sighted and has very small eyes

18

Lesser White-toothed Shrew

(Shrew Family)   *Crocidura suaveolens*

Identification: HB 5.5-8 cm T 2.5-4 cm W 3-7 g. Very similar to Greater
White-toothed Shrew (p. 18) but smaller; ears large, leathery and obvious;
tail with individual pale bristles.

Distribution: Central and southern Europe, central and southern Asia, North
Africa. In the British Isles found only in the Channel Isles (Sark and Jersey)
and most of the Scilly Isles.

Habitat: Warm places in open cultivated countryside, steppe landscape,
gardens and buildings.

Behaviour: Most active at night, solitary, very lively. Grass nest either above
or below ground; female leads young in a caravan formation.

Food: Insects, worms, slugs, snails, young mice, carrion; rarely seeds and fruit.

Breeding: Mating season Mar-Sep; gestation 28 days; 2-4 litters per year, each
with 3-9 young. Development as for Bicoloured White-toothed Shrew (p. 20).

Mainly a Mediterranean species, rare north of the Alps, but found as far
north as Berlin and northern France. Often overlooked because of its
rather secretive habits but traces can be found in the pellets of birds of prey
or owls. Towards the north of its range mostly found close to rural
habitation where it can overwinter in buildings.

Often mistaken for the larger Greater White-toothed Shrew. The Lesser
White-toothed is rather more lively in its behaviour.

Pygmy White-toothed Shrew   *Suncus etruscus*

(Shrew Family)

Identification: HB 3.5-4.5 cm T 2.4-2.8 cm W 1.5-2 g. Coat uniform grey;
ears leathery and very large; tail long, with white bristles.

Distribution: Southern Europe, southern Asia, North Africa.

Habitat: Damp undergrowth, woods, gardens, agricultural fields; often
along stream sides.

Behaviour: Mainly nocturnal, solitary, climbs well.

Food: Spiders, small insects.

Breeding: Gestation 28 days; probably 3-5 litters per year, each with 2-5 blind,
naked young; suckled for 20 days, after which they are independent.

The Pygmy White-toothed
Shrew prefers to nest on sunny
slopes or between the stones of old
walls, in positions where the
warmth of the sun is retained. Its
small size means that it has a
perpetual battle against
temperature loss. In cold periods
or at times of food shortage this
species may go into a state of
torpor for several hours. During
this process its body temperature
is reduced and it is able to save
energy. This species is one of the
smallest mammals in the world.

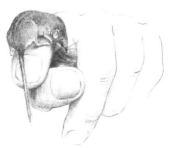

The Pygmy White-toothed Shrew is minute

Lesser White-toothed Shrew (above)
Pygmy White-toothed Shrew (below)

Bicoloured White-toothed Shrew

(Shrew Family)   *Crocidura leucodon*

Identification: HB 6.5-8.5 cm, T 3-4 cm, W 7-14 g. Dark grey-brown above, with sharp border to the white underside. Ears large and leathery; tail with long bristles.

Distribution: Central and southern Europe; southwest Asia. Not in the British Isles.

Habitat: Dry, sunny meadows; woodland edges; hedges and gardens.

Behaviour: Active day and night; solitary, but tolerant of members of same species. Less hectic in its movements than other shrews. Nests amongst grasses or below the ground in tunnels which it digs for itself, or in old mouse runs. The young occasionally form caravans behind their mother.

Food: Insects, snails, slugs, worms, other small invertebrates, carrion and sometimes fruit.

Breeding: Mating season Apr-Sep; gestation period 31 days; 2-4 litters per years, each with 3-9 naked blind young (birthweight about 1 g); young open their eyes at 13 days; suckled until 26 days and independent at about 40 days.

This species is easier to observe than other shrews because of its calmer, more trusting behaviour. The Bicoloured White-toothed Shrew makes a variety of shrill twittering cries in its communication with other individuals and is less aggressive towards members of its own species than other shrews. Male produces a strong musky scent from lateral flank glands, especially during the breeding season. If the female is ready for mating she will follow this scent.

The female of several white-toothed shrews leads her young in a caravan formation. One young shrew clamps its teeth firmly into the mother's fur and the other offspring form a line behind in a similar fashion. The entire family can thus be taken on a relatively safe expedition and quickly led away from any danger.

Bicoloured White-toothed Shrews in caravan formation

Pyrenean Desman   *Galemys pyrenaicus*

(Mole Family)

Identification: HB 11-14 cm, T 12-15 cm, W 50-80 g. Unique long flexible snout, expanded at tip. Tail-tip flattened from side to side. Hind feet webbed.

Distribution: Pyrenees, northwest Spain, northern Portugal.

Habitat: Fast-flowing streams and clean canals, mostly above 300 m.

Behaviour: Aquatic, foraging on the river bed but bringing food ashore to eat. Mainly nocturnal but some activity by day; solitary.

Food: Aquatic invertebrates, e.g. nymphs of stoneflies, worms, crustaceans.

Breeding: Mating season Jan-May, gestation period 4-5 weeks, litter of 1-4 born Feb-June, sometimes a second litter.

A related species, the **Russian Desman** (*Desmana moschata*), occurs in rivers of southern Russia. It is much larger, with a head–body length of 18-22 cm.

Female Bicoloured White-toothed Shrew with new-born young (above)
Pyrenean Desman (below)

Common Mole

Talpa europaea

(Mole Family)

Identification: HB 12-15 cm, T 2-4 cm, W 60-125 g. Body wedge-shaped; hair silky, dark grey to black; front feet directed sideways and shovel-shaped with long claws. Snout narrow and pointed; eyes tiny; external ears lacking.

Distribution: Whole of Europe except Ireland, Iceland and northern Scandinavia; western Asia. Widespread in Britain, replaced by very similar species in parts of the Mediterranean region.

Habitat: Loose soils, meadows, fields, gardens, deciduous woodland. Tends to avoid sandy, rocky or peaty soils.

Behaviour: Almost exclusively subterranean; active both by day and night. Strictly solitary except in breeding season; digs complex system of burrows with sleeping, nesting and living chambers. Soil extruded at the surface to form characteristic molehills.

Food: Mainly earthworms, also insects and their larvae, even occasionally shrews.

Breeding: Mating season Mar-May; gestation 30 days, 1-2 litters a year, each with 3-6 naked, blind young. Young open their eyes at about 3 weeks; suckled for 4-5 weeks and independent at about 2 months. Sexually mature in the following year.

Moles dig out their tunnels using their spade-like front limbs in a movement like breast-stroke, accumulating the loose soil behind them. From time to time the loose earth is pushed up to the surface through vertical side-tunnels. The soft, silky pelt enables the mole to move with equal ease forwards or backwards. Both the mouth and nostrils face downwards and are thus protected from the soil. The tiny eyes are also protected from roots and stones by the thick fur. The Mole's body contains many touch-sensitive hairs and bristles, not only on the

Moles are well adapted to their tunnel life, with a cylindrical body and short limbs

face but over the entire body, even on the tail. These help it in its search for prey through the long system of tunnels which may extend to up to 150 m in length.

In the autumn, Moles make a store of often hundreds of earthworms to last them through the winter. These stored worms are usually chewed off at the front end so that they remain alive but cannot crawl away. This provides a source of fresh food for months at a time. Moles eat the equivalent of their own body weight a day. They are usually strictly solitary, each with its own system of tunnels, and only come together during the breeding season. If territorially rival male Moles meet, fights often ensue in which the loser may be killed and might even be eaten.

Moles rarely come to the surface during the daytime

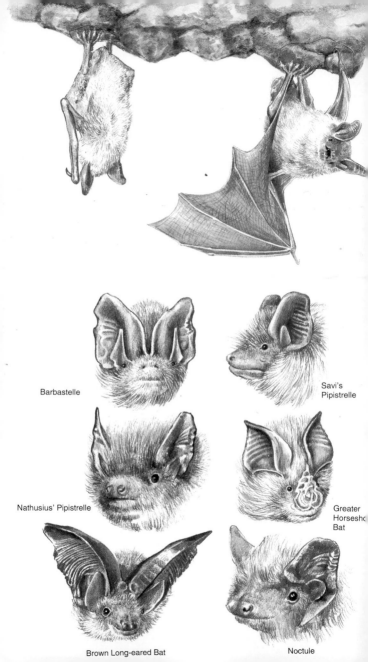

Barbastelle

Savi's
Pipistrelle

Nathusius' Pipistrelle

Greater
Horsesho
Bat

Brown Long-eared Bat

Noctule

Bats

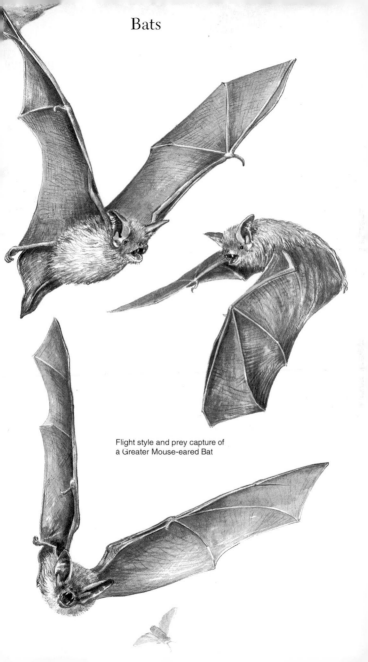

Flight style and prey capture of
a Greater Mouse-eared Bat

Greater Horseshoe Bat

(Horseshoe Bat Family)   *Rhinolophus ferrumequinum*

Identification: HB 5-6.5 cm, T 3-4 cm, WS about 37 cm, W 16-28 g. Fur grey to reddish-brown, creamy on the belly. Fleshy horseshoe-shaped nose. Ears large and pointed, lacking tragus (flap of skin at front of ear); wings broad.

Distribution: Central and southern Europe, North Africa, Asia. In Britain mainly in southwest England and southern Wales in precariously small numbers.

Habitat: Open country, scrub, light woodland.

Behaviour: Nocturnal; flight rather heavy and butterfly-like. Flies at height of 0.5-3 m; sociable. Summer roosts are in roofs, church towers and ruins. Hibernation Sep-Apr in caves, mines and cellars.

Food: Large flying insects, particularly moths.

Breeding: Mating season autumn and spring; gestation 10-11 weeks; a single blind, naked young (rarely twins); eyes open after a week and able to fly in 3-4 weeks; suckled for 6-8 weeks after which it is independent.

Note: Summer roosts are quite noisy.

The Greater Horseshoe Bat is a warmth-loving species. Its range extends north as far as the Harz mountains of Germany and southern England. It comes out around dusk and starts its hunting flights, which are usually fairly close above the ground. Orientation as in all bats is through echo-location and not by sight (although it is not blind). An uninterrupted stream of high-pitched calls is reflected off objects, including prey, and picked up as echoes by the bat's sensitive ears. These echoes give the bat information about both the size and distance of the object. These orientation calls are ultrasonic and can therefore only be detected by special instruments (bat detectors). Unlike the other European species, the Horseshoe Bats emit these noises through the nose. The strange fleshy nose appendages act as an acoustic focusing device. Bats also emit chirping and twittering sounds that we can hear; Greater Horseshoe Bats are particularly noisy in their summer roosts. The young are also noisy when waiting to be fed.

Bats are the only mammals capable of active flight. The bat's wing is a remarkable piece of adaptation, involving the limbs and tail. The actual surface of the wing consists of a thin layer of skin stretched between the enormously long lower arm, hand and finger bones. The legs and tail are also involved. Running through this membrane are blood vessels and also elastic threads. The latter help to pull the wing together when it is folded. Hibernating bats wrap themselves up almost completely in their wings.

To encourage bats in your garden special bat nest boxes can be installed where old trees are rare.

Hibernating Horseshoe Bats hang from the ceiling and maintain a distance between each other

Greater Horseshoe Bat in flight

Lesser Horseshoe Bat
Rhinolophus hipposideros

(Horseshoe Bat Family)

Identification: HB 3.5-4.5 cm, T 2.4-3 cm, WS about 20 cm, W 4-10 g.
Similar to Greater Horseshoe Bat (p. 28), but much smaller; fur
grey-brown and whitish-grey beneath; fleshy nose; ears pointed, lacking
tragus.

Distribution: Central and southern Europe, North Africa, Southeast Asia. The
most northerly horseshoe bat, reaching to the Mosel Valley in Germany. Also
found in the west of Ireland the west and southwest of Britain.

Habitat: Open woods, parks.

Behaviour: Nocturnal; hunting flight begins late at night and ends before
dawn. Flight slow, 2-5 m above the ground. Sociable species; summer
roosts often in large attics. Males usually solitary in summer. Hibernation
Sep-May in caves, mines and cellars.

Food: Small flying insects.

Breeding: Mating season autumn and spring; gestation about 10 weeks with
a single naked blind young (rarely twins); eyes open after a week; flies after
3-4 weeks; suckled 6-8 weeks, after which independent.

This species is less sensitive to cold than its relatives. It seeks out deep
caves or mines in which to hibernate and requires a temperature of not less
than 5-9 °C. Individuals are faithful to particular hibernating sites.
Hibernates almost entirely shrouded in its wing membranes, often with
only the tip of its nose protruding
and the tail pressed against the
back. Three other species of
horseshoe bats occur in parts of the
Mediterranean region: **Blasius'
Horseshoe Bat** (*Rhinolophus
blasii*), **Mehely's Horseshoe Bat**
(*Rhinolophus mehelyi*) and
Mediterranean Horseshoe Bat
(*Rhinolophus euryale*).

Nose shapes in Horseshoe Bats: (left to
right) Greater, Lesser and Mediterranean

Barbastelle
Barbastella barbastellus

(Vesper Bat Family)

Identification: HB 4.5-5.8 cm, T 3.8-5.4 cm, WS 24-29 cm, W 6-13 g.
A small, dark bat, with short rounded ears that meet on top of the head, a
condition found elsewhere only in the long-eared bats.

Distribution: Europe from southern Scandinavia to the Mediterranean, east
to the Caucasus. Widespread but very local in England and Wales.

Habitat: Mainly woodland.

Behaviour: Very elusive, especially in summer when roosts in small
numbers in crevices in trees. Emerges to fly around at sunset. Hibernates
Oct-Apr, usually underground.

Food: Small insects, often caught over water or picked from foliage.

Breeding: Mating in autumn and winter. A single young is born in June.

Lesser Horseshoe Bat hibernating (above left)
Mediterranean Horseshoe Bat (above right)
Barbastelle (below)

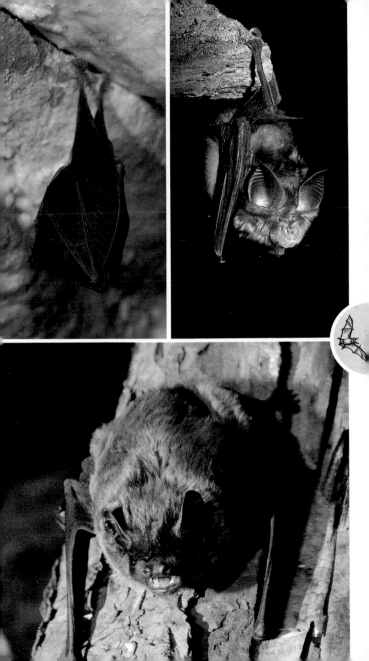

Greater Mouse-eared Bat

Myotis myotis

(Vesper Bat Family)

Identification: HB 6.5-8 cm, T 4.5-6 cm, WS to 40 cm, W 20-40 g. Fur grey-brown above, lightish grey beneath; wings broad and dark grey. Ears broad and oval, tragus reaching just under half-way to tip of ear. Summer roosts noisy (shrill scolding and shrieking).

Distribution: Southern and central Europe, Ukraine, Turkey, rather rare in northern and western Europe, absent from Ireland and recently extinct in England although there were never more than a few scattered colonies in southern England.

Habitat: Open river valley woods, parks, fields and meadows.

Behaviour: Flies soon after dusk and only in good weather; flight slow, straight and about 5-8 m high. Summer roosts in large dark buildings or tree holes. Hibernates individually or in small groups in caves, mines and cellars.

Food: Large insects caught in flight (moths) or on the ground (beetles, crickets).

Breeding: Mates in autumn and spring; gestation 50-60 days; a single naked blind young, capable of flight at about 40-50 days old and independent at about 2 months.

This is one of our largest native bats. The males of this species often roost individually, but the females tend to live in colonies, sometimes of up to several hundred. The young are born in these female colonies. In autumn they seek out dark, frost-free caves with high humidity, and these can be as much as 200 km from their summer quarters.

The slightly smaller **Lesser Mouse-eared Bat** (*Myotis blythii*) occurs throughout the Mediterranean region.

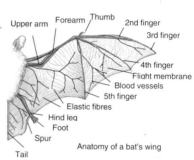

Anatomy of a bat's wing

Bechstein's Bat

Myotis bechsteinii

(Vesper Bat Family)

Identification: HB 4.3-5.5 cm, T 3.4-4.4 cm, WS 25-30 cm, W 7-13 g. Fur reddish-brown above, pale grey below. Ears very long (2-2.5 cm) but shorter than in long-eared bats and very clearly separated from each other.

Distribution: Europe from southern Scandinavia to the Mediterranean, east to the Caucasus. Very local in southern England.

Habitat: Woodland.

Behaviour: Emerges after sunset and hunts amongst trees. Roosts in small numbers in tree holes; very difficult to locate.

Breeding: A single young born in June or July.

Greater Mouse-eared Bats hang on to
the ceiling with their hind legs (above left)
Greater Mouse-eared Bat in flight (above right)
Bechstein's Bat (below)

Daubenton's Bat

Myotis daubentonii

(Vesper Bat Family)

Identification: HB 4-5.3 cm, T 3-4 cm, WS 21-25 cm, W 6-12 g. Fur grey-brown, pale grey beneath; hind legs very large; tip of tail extends very slightly beyond flight membrane; ears oval, tragus small and pointed.

Distribution: Whole of Europe except northern Scandinavia and Iceland, rare in southeast Europe; Asia. Widespread in Britain and Ireland.

Habitat: Wooded country with large stretches of open water, lakes and river banks; also cultivated areas.

Behaviour: Active at dusk and by night, often hunts over the water surface; flies with very rapid wing beats, usually about 2 m above the ground. Summer roosts in hollow trees and crevices in buildings; female colonies of over a hundred; hibernates Oct-Mar in rock crevices or cellars.

Food: Midges, gnats, moths.

Breeding: Mates from autumn to spring; gestation 50-60 days; one young per year, independent after 6-7 weeks.

This species is characteristically seen flying in a regular circuit over the water surface. Small insects such as midges are eaten in flight, larger ones are taken back to a tree branch. Daubenton's Bats tend to use the same hole for hibernation every year. There they sit horizontally on stones and in crevices. Disturbances during hibernation can be fatal, as each time the bats are woken they raise their body temperature, thus reducing vital fat reserves.

Three other very similar species occur in western Europe but not in the British Isles: **Long-fingered Bat** (*Myotis cappaccinii*), **Pond Bat** (*Myotis dasycneme*) and **Geoffroy's Bat** (*Myotis emarginatus*).

Natterer's Bat

Myotis nattereri

(Vesper Bat Family)

Identification: HB 4.2-5 cm, T 3.2-4.3 cm, WS 22-27 cm, W 6-12 g. Fur grey-brown, whitish beneath; face, ears and wing membrane grey-brown; stiff, curved bristles around the edge of tail membrane; ears large and oval, tragus long and narrow.

Distribution: All Europe except northern Scandinavia and the Balkans; North Africa; Asia. Widespread but sparse in Britain and Ireland.

Habitat: Open woods, parks, and gardens; often near water.

Behaviour: Active at dusk and by night; flight slow, whirring, regular and straight. Sociable, usually hunts in small groups. Summer roosts in hollow trees, nest-boxes, sometimes in holes in walls. They hibernate individually or in small groups in damp crevices or damp caves or cellars.

Food: Small flying insects such as small flies, midges, small moths.

Breeding: Similar to Daubenton's Bat.

Natterer's Bats often hunt along hedgerows and in orchards, where they take small insects from the twigs. The increasing use of insecticide sprays has reduced their food. They can also be poisoned after consuming insects that have taken in pesticides. To protect and encourage bats do not use chemicals in your garden.

Whiskered Bat (*Myotis mystacinus*) and **Brandt's Bat** (*Myotis brandtii*) are similar, but are more greyish, with all the naked skin almost black. Both are widespread in England and Wales (Whiskered Bat also in Ireland) but are difficult to distinguish.

Daubenton's Bat (above)
Natterer's Bat (middle); Whiskered Bat (bottom)

Common Pipistrelle

Pipistrellus pipistrellus

(Vesper Bat Family)

Identification: HB 3.3-5 cm, T 2.5-3.5 cm, WS to 19 cm; W 3-8 g. Fur colour very variable, from dark blackish-brown to yellow or rusty brown, usually grey-brown coat underneath; face, ears and wing membranes dark grey; ears and tragus short and rounded; wings narrow.

Distribution: Whole of Europe except northern Scandinavia, northern Russia and Iceland; northwest Africa; western and central Asia. The most widespread species of bat in Britain and Ireland.

Habitat: Open country with isolated trees; open woodland, gardens, parks; also found in towns and cities, and in mountains right up to the tree line.

Behaviour: Active at dusk and by night; flight rapid, with a lot of twisting; flight height 2-6 m; very sociable; summer and winter roosts in narrow bark or rock crevices, often in buildings. Hibernates Oct-Mar, usually in rock crevices.

Food: Small flying insects such as midges, gnats and moths.

Breeding: Mating season Apr-May; gestation 4-6 weeks; 1 or 2 naked blind young in each litter; young fly after 3 weeks and are independent at 2 months.

Note: Although this species is mainly sedentary, some long migrations of up to about 800 km have been recorded.

The Common Pipistrelle is the smallest European bat species and also our commonest. It can be seen in villages, towns and even in the centres of large cities. They begin their hunting flights in the early evening and can even be seen during the afternoon on mild winter days, and sometimes even in cool and rainy weather. They often cruise around street lamps, feeding off the insects which are attracted to the light. Common Pipistrelles often do not return to their roosts until the first light of dawn.

Favourite roosting sites are narrow gaps beneath loose tree bark, behind window shutters, behind panels and in wall crevices. Pregnant females often gather in roof spaces, where the young are born in June. Such nursery roosts can contain several thousand animals.

The new-born Common Pipistrelle weighs only about 1 g, but like all bats they grow rapidly and can fly after three weeks. They quickly learn from their mothers how to catch insects but still depend on her for milk for the first couple of months. After this time the young leave the nursery roosts and become independent, and at this stage the adult females start to associate with the males again.

Four-day-old Common Pipistrelle

Dew forming on the fur of a sleeping Common Pipistrelle

Savi's Pipistrelle

Pipistrellus savii

(Vesper Bat Family)

Identification: HB 4.5-4.8 cm, T 3.4-3.9 cm, WS 22-23 cm, W 6-10 g. Similar to Common Pipistrelle (p. 36) but slightly larger. Fur blackish-brown, often with lighter tips to hairs; face, ears and wing membranes black; tip of tail extends clearly beyond tail membrane.

Distribution: Southern Europe, northwest Africa, Asia. A single record in southern England.

Habitat: Mountain pastures and valleys; wooded areas and cultivated land.

Behaviour: Active at dusk and night; flight slow, regular; in summer groups of up to 30 are formed; summer roosts in rock crevices, hollow trees, mountain huts; usually hibernates individually in narrow crevices in hollow trees, walls or rocks.

Food: Small flying insects.

Breeding: As for Nathusius' Pipistrelle.

During hibernation the body temperature of all bats drops from about 40 °c down to near the temperature of the surroundings. In bad weather bats become very sluggish, and Savi's Pipistrelle is no exception. When too cool they are incapable of flight and can only move very slowly. Unlike so-called cold-blooded animals (such as reptiles and amphibians) sleeping bats are able to raise their body temperature again. Each evening before the hunting flight, and also if disturbed mid-hibernation, bats can warm themselves up to their normal operating temperature in just a few minutes. During torpor the bat's energy requirement is drastically reduced.

Nathusius' Pipistrelle

Pipistrellus nathusii

(Vesper Bat Family)

Identification: HB 4.6-5.4 cm, T 3.3-4 cm, WS 23-24 cm, W 6-12 g. Resembles Common Pipistrelle (p. 36), but slightly larger; fur brown, greyish-brown underneath; often with a darker shoulder spot; face, ears and wing membranes brown.

Distribution: Central and eastern Europe, Turkey. Widespread in Netherlands; recorded rarely in southeast England and the Channel Islands.

Habitat: Open deciduous and mixed woodland; also parks and gardens.

Behaviour: Active at dusk and by night; flight rapid and twisting at a height of 5-15 m; lives in small colonies; summer and winter roosts are in tree holes, wall or rock crevices.

Food: Small flying insects.

Breeding: Mating season Apr-May; gestation 6-8 weeks; 1 or 2 naked blind young which fly after about 3 weeks and are independent at 2 months.

Like all bats, Nathusius' Pipistrelle spends a lot of time keeping its body in a good condition. When a bat lands it often combs its coat repeatedly using its teeth and a hind foot. Special glands at the corners of the mouth secrete an oil which keeps their wing membranes lubricated. They rub their half-opened wings across their face to smear this secretion across the membrane. This protects the wings from drying out. Nathusius' Pipistrelle is a migratory species. Those breeding in northeast Europe move south west in autumn, and there are several instances of movements of well over 1000 km.

One further species of Pipistrelle, **Kuhl's Pipistrelle** (*Pipistrellus kuhlii*) occurs in the Mediterranean region.

Savi's Pipistrelle (above)
Nathusius' Pipistrelle on hunting flight (below)

Noctule

(Vesper Bat Family)

Identification: HB 7-8 cm, T 4-5.5 cm, WS 33-46 cm, W 15-40 g. Fur yellowish or reddish-brown, slightly paler beneath; face, ears and wing membrane blackish-brown; ears and tragus short and round; wings long and narrow.

Distribution: Whole of Europe except northern Scandinavia, Iceland, Ireland and much of Scotland.

Habitat: Deciduous and mixed woodland; parks and gardens.

Behaviour: Active at dusk and night; sometimes also seen during the daytime. Flight rapid and straight with quick turns and dives, usually quite high up and often above the trees (up to 80 m); sociable; males and females live apart during the summer; summer roosts in hollow trees, nest boxes, cracks in walls; females gather in nursery roosts, often with more than 100 together; hibernation Oct-Apr in thick clusters of up to 1000 animals in hollow trees or roof spaces.

Food: Flying insects.

Breeding: Mate in autumn and spring; gestation 6-8 weeks; 1 or 2 naked, blind young, suckled for 4-6 weeks, after which they fly and become independent.

The Noctule is often seen as early as late afternoon, flying high in search of insect prey. The flight is straight and rapid, often interrupted by sudden pauses as it catches its prey. The narrow, swallow-like wings often almost touch under the body during flight and the irregular alternation between deep and shallower wing strokes is a good identification character for this species. Wings beat seven or eight times a second, producing a top speed of about 50 km per hour.

The young are born in June. When giving birth, the female positions herself horizontally, with the head facing slightly upwards, and bends her tail forwards toward the belly, forming a pocket into which the new-born young emerges. The young bat then clambers into the mother's fur in search of one of the two nipples. At this stage it may still be attached by the umbilical cord. It then clamps itself tight on the nipple using its teeth. In the evening when the mother emerges to hunt the young release themselves and cling on tightly to the wall of the roost waiting patiently for her return. Each female tends only her own young which she can distinguish from the many others by their particular voice and smell. The Noctule is a migratory species, travelling up to 1500 km between summer and winter quarters.

Leisler's Bat

(Vesper Bat Family)

Identification: HB 5-7 cm, T 3.5-5 cm, W 11-20 g. Like Noctule but smaller and fur less rufous, dark brown when parted.

Distribution: Central and southern Europe; east to the Himalayas; scattered in England and Wales, more widespread in Ireland.

Habitat: Woodland.

Behaviour: As Noctule but more often lives in buildings.

Other details as Noctule.

Mother Noctule with young (above left)
Noctule emerging from hollow tree (above right)
Leisler's Bat (below)

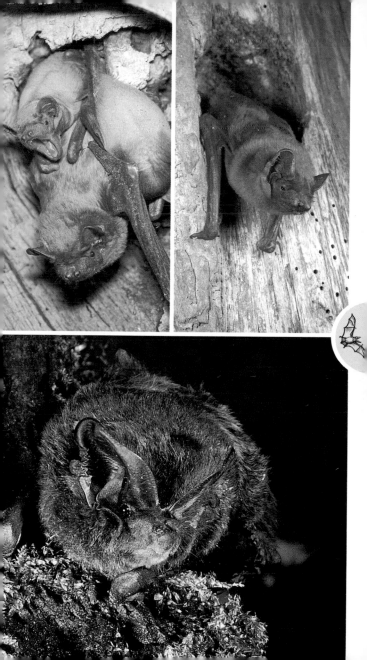

Serotine Bat

Eptesicus serotinus

(Vesper Bat Family)

Identification: HB 6.2-8 cm, T 4.6-6 cm, WS 34-36 cm, W 17-35 g. Fur dark brown, yellowish-brown beneath; face, ears and wing membranes blackish-brown; ears short and wide, with a small tragus; wings broad; tip of tail extending about 6 mm beyond membrane.

Distribution: Central and southern Europe, but also north to Denmark, southern Sweden, southern Wales and southern England; northwest Africa; Asia.

Habitat: Open country with groups of trees, often near habitation.

Behaviour: Nocturnal. Flight slow, at a height of 6-10 m. Summer roosts in crevices in rocks or walls, also in hollow trees; nursery roosts of up to 50 females; males solitary in summer. Hibernates in rock clefts or caves.

Food: Large flying insects; also ground insects.

Breeding: Mates in autumn and spring; gestation 6-8 weeks; single naked, blind young (rarely twins) a year; young flies at about 4-6 weeks, suckled for 2 months and is then independent.

This warmth-loving species is most common in the milder parts of Europe. Hunting flights do not usually start until it is quite dark. Serotines sometimes hover in flight in order to pluck insects or spiders from twigs or walls, or to take caterpillars from leaves. They can also walk quite well on the ground, as can other bats. When walking, the bat raises itself with the lower arms then heaves the body forward with the hind legs. It then takes off again by jumping straight up from the ground, just high enough that it can spread its wings again.

Bats rarely enter water willingly. If they do fall into water they try to row themselves with wing strokes towards the nearest bank. In Europe there are no predators that specialise particularly on bat prey, but owls and birds of prey do occasionally take bats. Periods of bad weather or sudden frosts are of much greater danger to bats. In rainy summers there is often a scarcity of insect prey so bats may starve or build up insufficient reserves to see them through hibernation.

The related **Northern Bat** (*Eptesicus nilssonii*) is a migrant species that reaches north of the Arctic Circle in summer. It has been recorded once in the south of England.

Daubenton's Bat

Serotine

The number of folds in the ear and the shape of the tragus are good characters for bat identification

Noctule

Parti-coloured Bat

Serotine Bat disturbed at its roost under a roof

Brown Long-eared Bat

Plecotus auritus

(Vesper Bat Family)

Identification: HB 4-5.2 cm, T 3.8-5 cm, WS 22-26 cm, W 5-11 g. Hair brown or grey-brown, whitish-grey beneath; ears very large and meeting in the middle; tragus long and pointed, translucent; wings broad.

Distribution: Europe, except northern Scandinavia, Iceland, Iberian Peninsula, southern Italy and Mediterranean Islands. Temperate parts of Asia. Britain's second most common bat.

Habitat: Open woodland scrub; parks and gardens.

Behaviour: Active at dusk and by night; flight slow; not very sociable; nursery roosts of 10-20 animals, including some males; summer roosts in hollow trees, nest boxes, also in buildings; hibernation Oct-Apr, usually in caves or cellars.

Food: Moths, beetles, caterpillars, midges.

Breeding: Mating season autumn and spring; gestation 6-8 weeks; single naked, blind young per year (rarely twins); suckled for 6 weeks, after which independent; sexually mature in 2-3 years.

When at rest a Long-eared Bat keeps its ears folded back and extends them just before flying off. When hibernating they are folded backwards and also tucked under the wings.

Like most of our bats, Long-eared Bats indulge in autumnal display flights. They chase each other backwards and forwards, making audible 'tzick-tzick' calls. At this stage most of the mature females pair-up with the males. They are however not yet fertile since their egg cells only mature the following spring. The sperm from the male bats remains viable in the female's body throughout the winter, until after hibernation when her eggs are released and fertilisation takes place. Females which did not mate the previous autumn do so early in the following year and become pregnant without this delay mechanism. Since bats adjust their metabolic rate to the ambient temperature while they sleep, the gestation period is very dependent upon the weather. In bad weather this may be considerably extended, and it is similarly decreased during warm conditions.

The similar **Grey Long-eared Bat** (*Plecotus austriacus*) was only recognised as a distinct species as recently as 1960. It is widespread in central and southern Europe, with a few colonies on the south coast of England. Although very similar to the Brown Long-eared Bat, in biology as well as appearance, these two species seldom compete since they are rarely found together in the same region.

Long-eared Bat at rest, showing
ears folded back under the wings

Brown Long-eared Bat (top)
Grey Long-eared Bat (bottom)

Parti-coloured Bat
Vespertilio murinus

(Vesper Bat Family)

Identification: HB 5.5-6.5 cm, T 4-4.5 cm, WS 26-30 cm, W 12-18 g. Fur black, tipped silvery grey; whitish beneath; face, ears and wing membranes brownish-black; ears rounded; wings narrow; tip of tail extends beyond tail membrane.

Distribution: Central and eastern Europe, north to southern Scandinavia, scattered in western and southern Europe; Asia. Migrates from northern to southwest Europe and occurs in British Isles only as a vagrant.

Habitat: Montane forests; mixed agricultural areas; villages; towns.

Behaviour: Nocturnal; flight rapid, at a height of about 20 m; summer roosts in clefts, often on tall buildings, also in hollow trees; males roost in groups of up to 100, females in nursery roosts in small groups or individually; hibernates Oct-Mar in large colonies in rock crevices, caves or cellars.

Food: Large insects.

Breeding: Similar to Serotine (p. 42), but usually 2 young per litter.

This species, like all narrow-winged bats, has a rapid and agile hunting flight. Once it has detected a flying insect it is almost always successful. Certain insects however have evolved sophisticated defence mechanisms. Particular moths drop like a stone as soon as they detect the ultrasound call of a hunting bat. Other moths have furry bodies which help to damp down the echoes and thus serve to confuse the bat.

Unlike other bats, female Parti-coloured Bats tend to rear their young alone rather than in nurseries.

European Free-tailed Bat
Tadarida teniotis

(Free-tailed Bat Family)

Identification: HB 8.2-8.7 cm, T 4.5-5.5 cm, WS about 40 cm, W 25-50 g. Fur dark grey to grey-brown; face, ears and wing membrane blackish-grey; ears large and meeting across the forehead; tragus small; wings long and very narrow; about a third of the tail extends beyond the tail membrane.

Distribution: Mediterranean Europe and North Africa; temperate parts of Asia.

Habitat: Rocky areas, cliffs, bridges, tall buildings.

Behaviour: Active at dusk and by night; often flies with swallows or swifts; flight straight, rapid and high; not very sociable; summer and nursery roosts in rocks and walls; short or intermittent hibernation usually in caves.

Food: Large flying insects.

Breeding: Mating season autumn and spring; gestation period 75-85 days; single naked, blind young per year which flies after 3-4 weeks, suckled for 7-8 weeks and is then independent.

The European Free-tailed Bat often flies in the late afternoon. In addition to the ultra-sound calls this species also has a clearly audible flight call. The long, narrow wings give it a rapid powerful flight similar to a swallow. Although fast, its flight is silent, except for the occasional rustle as it banks or changes direction.

Parti-coloured Bat (above)
European Free-tailed Bat (below)

Barbary Ape

Macaca sylvanus

(Old-world Monkey Family)

Identification: HB 60-75 cm; T lacking, W 5-10 kg. Appearance and body shape typically ape-like; coat thick and ochre brown; tail not visible; female displays swollen buttocks when on heat.

Distribution: Gibraltar, northwest Africa.

Habitat: Rocky country with scrub or open woodland.

Behaviour: Diurnal; family groups of between 15-30 animals led by a dominant male; usually travel on the ground, climbing only in order to sleep or to escape from danger; jumps well; young cling on to their mother's fur.

Food: Omnivorous: grain, herbs, fruit, seeds, insects, small vertebrates.

Breeding: Mating season throughout the year; gestation 6 months; a single young with thin black hair at first; suckled until 3 months old and independent at one year.

The Barbary Apes of the Rock of Gibraltar are the only free-living monkeys in Europe. It is still not certain whether these were introduced deliberately or whether they represent the remains of an original south European population. It is most likely however that the Gibraltar colony originated from north African stock and it is now artificially maintained at the level of about 40 animals.

Females have a four-week sexual cycle. When they are receptive this is signalled by a significant swelling of the skin around the sexual organs. New-born offspring cling to the mother's belly fur and are carried around in this manner. At a week old they can walk for themselves, but tend to ride on their mother's back over long distances. Young subordinate males often help to carry young monkeys around. It seems this activity also protects them from attack by the dominant male of the troop. They also indulge in mutual grooming, and this plays an important role in their family life. They are not usually looking for fleas but are generally involved in removing dead skin and keeping their hair clean. They also signal mutual trust to each other through this activity and this strengthens the social ties of the group. Barbary Apes communicate with one another using a series of distinct calls.

This yawn is actually a threat gesture
and reveals the sharp canine teeth

Mother with young baby (above)
A grooming session (below)

Mice, mouse-like rodents

Northern Birch Mouse

Striped Field Mouse

Spiny Mouse

Harvest Mouse

House Mouse

Bank Vole

Black Rat

Wood Mouse

Northern Water Vole

Common Pine Vole

Common Vole

Common Rat

Snow Vole

and shrews

Forest Dormouse

Fat Dormouse

Common Dormouse

Garden Dormouse

Shrews (Insectivores)

Bicoloured White-toothed Shrew

Water Shrew

Common Hamster

Greater White-toothed Shrew

Pygmy Shrew

Pygmy White-toothed Shrew

Lesser White-toothed Shrew

Alpine Shrew

Norway Lemming

Common Shrew

Red Squirrel

Sciurus vulgaris

(Squirrel Family)

Identification: HB 20-25 cm, T 15-20 cm, W 230-450 g. Colour of coat very variable, from foxy-red (the usual form in Britain) through yellowish or dark brown to almost black in some areas; tail bushy; ears with long tufts of hair in the winter; hind legs longer and more powerful than front legs; makes a loud 'chuck-chuck' noise when alarmed.

Distribution: The whole of Europe except Iceland and the islands of the Mediterranean. Still widespread in much of Scotland and Ireland; very scattered in England and Wales, where it has been replaced by Grey Squirrel (see p. 54).

Habitat: Deciduous, mixed and coniferous woodland; parks, gardens.

Behaviour: Diurnal; climbs and jumps very well; almost entirely tree-living; usually solitary; builds a twig nest in a fork of a tree, sometimes inside a hollow tree or in an old crow's nest; male and female chase each other during the mating season; collects food and stores it in hollow trees or buried in the ground; does not hibernate.

Food: Seeds, nuts, berries, fungi, bark, buds, shoots; also insects, birds' eggs and nestlings.

Breeding: Mating season Dec-June; gestation about 38 days; 1-2 litters a year, each with 2-5 naked, blind young; these open their eyes at about a month, are suckled for 2 months, after which they are independent.

The Red Squirrel is superbly adapted for tree-climbing. Its sharp, curved claws enable it to get a firm grip on bark and it can even descend a tree trunk head-first. The bushy tail lends stability when the squirrel jumps from branch to branch and also acts to some extent as a brake. Squirrels do not usually gain much height when jumping as they usually clamber as far as the thinnest twigs before taking off. Pine Martens are the only predators that can successfully catch squirrels in the tree-tops.

Red Squirrels are mainly vegetarian, although they do occasionally plunder birds' nests. Like all rodents, they have powerful incisor teeth – one pair each in the upper and lower jaw. These teeth are very long and curved, and grow throughout the life of the animal, being worn down through perpetual gnawing. The dentine comprising the bulk of the tooth wears down more quickly than the enamel at the front of the tooth, so that the teeth develop a sharp chisel-like profile. In rodents, both the canine and some or all of the pre-molar teeth are absent, leaving a gap (diastema) between the incisors and the molars at the back of the mouth. The molars are flat and are used to grind up the food. Like almost all rodents, squirrels have a split upper lip (hare lip), through which the upper incisors can protrude when gnawing food. (Continued p. 54.)

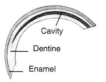

Cavity

Dentine

Enamel

Cross-section of squirrel skull (above) and incisor tooth (below)

This young squirrel all ready has the ear-tufts characteristic of the winter coat

Red Squirrel (continued)

Squirrels build quite elaborate twig nests – usually one main nest and one or more subsidiary nests. These are usually anchored in a fork in the branches high in a tree. Squirrel nests or dreys can be distinguished from a large bird nest because the entrance hole is on the underside. The drey is lined with grass and moss and is used both to rear the young and as a resting place during the winter.

Squirrels do not hibernate in the strict sense, but show a form of dormancy rather than true hibernation. In cold weather they are much less active and sleep for longer periods, often days at a time. Squirrels use their excellent sense of smell to help them find seeds and nuts which they stored the previous autumn. A proportion of the stored nuts are often not found and eaten and in this way squirrels help in the dispersal of tree seeds in woodland.

Interior of a drey with naked blind young

The characteristic ear tufts of the Red Squirrel are best developed during the winter. These tufts are lost during the spring moult and do not redevelop until the following winter. This species shows much geographical variation in coat colour. For instance, in the high Alps the commonest form of Red Squirrel is dark blackish-brown, but in many lowland areas including the United Kingdom it is bright reddish-chestnut. Indeed in some places more than one colour form may be found at a single locality.

Grey Squirrel
Sciurus carolinensis
(Squirrel family)

Identification: HB 23-30 cm, T 20-24 cm, W 450-650 g. Coat grey above, white below, with a variable tinge of orange-brown especially on the flanks. Ears always untufted.

Distribution: A North American species, now widespread in lowland Britain and Ireland where it has replaced the Red Squirrel in many areas; also introduced to northern Italy.

Habitat: Broad-leaved and mixed broad-leaved/coniferous woodland; suburban parks and gardens.

Behaviour: Diurnal; most time spent in trees but comes to ground more frequently than Red Squirrel. Nest usually conspicuous in crown of tree.

Food: As for Red Squirrel; acorns and hazel nuts are often predominent, but they also open unripe cones to extract seeds.

Breeding: Gestation 42-45 days. One litter of 2-4 young in spring, occasionally a second. Young weaned at 10 weeks.

Grey Squirrels have replaced the native Red Squirrels in much of Britain, probably by competition for food during the harshest part of winter rather than by direct aggression. Red Squirrels appear to retain the advantage only in stands of pure conifers.

Red Squirrel in summer coat (above)
Grey Squirrel (below)

Flying Squirrel

Pteromys volans

(Squirrel Family)

Identification: HB 15-18 cm, T 10-13 cm, W 130-200 g. Resembles Fat Dormouse (p. 86) when sitting; summer fur yellowish-grey, in winter silver-grey, white beneath; 'wings' formed by stretched skin between limbs when jumping. Tail bushy and rather flattened.

Distribution: Northeast Europe and Asia, Finland and the Baltic through Siberia and eastwards.

Habitat: Deciduous and mixed woodland, particularly birch and conifers.

Behaviour: Mainly active at dawn and dusk or by night; agile climber; glides from tree to tree; nests in hollow trees or old woodpecker holes, squirrel dreys or abandoned birds nests; does not hibernate; occasionally stores food; threatened by disturbance and forestry.

A highly distinctive species particularly when gliding, when it uses the stretched skin as a kind of parachute. This enables it to extend its leaps to beyond 30 m; it can thus escape predators such as Pine Martens. Unlike a bat, a Flying Squirrel is not capable of active flight.

Flying Squirrel gliding

Siberian Chipmunk

Tamias sibiricus

(Squirrel Family)

Identification: HB 13-15 cm, T 8-10 cm, W 50-120 g. Resembles a squirrel, but markedly smaller; fur yellow-brown, white belly; five dark brown longitudinal stripes along the back; tail bushy and striped.

Distribution: Northern Europe and Asia from Finland to eastern Siberia. Some colonies have been established from escapes in France, Germany, Netherlands, Belgium and Austria.

Habitat: Coniferous and mixed woodland with thick undergrowth; woodland margins and river valleys.

Behaviour: Diurnal; climbs well; clasps food firmly in its front paws when feeding; colonial; digs extensive burrows in the soil; hibernation Oct-Mar or Apr.

Food: Seeds, berries, fruits, fungi, buds, insects.

Breeding: Mating season Apr-May; gestation 5-6 weeks; 1 or 2 litters a year, each with 4-6 (up to 10 maximum) naked, blind young; young suckled for 4 weeks and independent at 4 months.

Lives colonially in adjacent burrow systems; each individual defends its own territory and marks the borders with urine. From August begins to store provisions for the winter by collecting seeds, nuts and berries and transporting these back to the nest in cheek pouches like a hamster. Each animal stores up to about 5 kg in this way.

In some parts of the former Soviet Union, Brown Bears are a major predator during the winter. The bears dig out the burrows and then consume both the stored food and the Chipmunks themselves.

Flying Squirrel (above)
Siberian Chipmunk (below)

Alpine Marmot

Marmota marmota

(Squirrel Family)

Identification: HB 47-60 cm, T 13-20 cm, W 3-8 kg. Compact body; fur grey to yellowish-brown; ears small, almost covered by fur; legs short and powerful; claws short; tail with bushy hairs; can emit a shrill alarm call.

Distribution: Central and western Alps and the Tatra mountains, Slovakia; recently introduced to Pyrenees, Carpathians, Black Forest and eastern Alps.

Habitat: Alpine grassland above the tree line, usually 1500-3200 m; sometimes at lower altitudes on alpine meadows.

Behaviour: Diurnal; lives in family groups in colonies; marks borders of territory with scent from cheek glands; constructs complicated system of burrows for the winter, also shallower summer burrows and simple bolt-holes; moves with a rolling gait; often sits up bolt upright; hibernation Oct-Apr.

Food: Grass, herbs, roots, berries.

Breeding: Mating season Apr-May; gestation 34-35 days; usually 2-3 (up to 7) naked, blind young; these open their eyes at about 3 weeks old, are suckled for 2 months and are sexually mature at 3 years.

Alpine Marmots are typically found on high alpine meadows and screes. In some places they also colonise meadows below the tree line. The small family groups consist of an original pair together with offspring of varying age. Most young animals remain in the family until their third year, at which stage they generally wander off in search of a mate. Each family defends a territory of about two or three hectares. Only very young animals are tolerated in neighbouring territories, where they may play together with other young marmots.

It is usual for many family groups to live close together in a colony, and in this case the myriad burrows sometimes undermine entire hill slopes. The larger winter quarters may go as deep as 3 m into the ground.

Cross-section through burrow system

At the very bottom of the Alpine Marmot burrow system is a chamber lined with hay. Marmots do their own hay-making. In late summer they bite off long grass stems and lay them in the sun to dry. When the hay is ready it is collected, either in cheek pouches or in the mouth, and taken underground. The marmots also make underground stores of food in readiness for early the following year when fresh grass is scarce. The whole marmot family goes underground, often even before the first snow falls in the autumn. They seal the entrances carefully using soil and hay and hibernate close together in their underground chambers. (Continued p. 60.)

Alpine Marmot making alarm call (above)
Alpine Marmots become slimmer after hibernation (below)

Alpine Marmot (continued)

As with all hibernating mammals their body temperature sinks to a minimum level after a few days, and their heart beat is reduced from the normal waking rate of 90-140 times a minute to a mere three to five times a minute.

Similarly the breathing rate also drops and body temperature falls to 4-7 °C. In this way the energy requirement is significantly reduced, enabling the marmot to survive through the winter months on its body fat reserves. They do however wake every two to three weeks mainly in order to pass faeces and urine but they don't actually feed at this stage. If the temperature drops too far the animal will wake before it freezes to death. By sleeping with their bodies close together they lose less heat than they would lying individually. Young animals in particular, with their smaller fat reserves, are more likely to survive the winter when hibernating in groups.

When they emerge the following spring as the snow melts the Alpine Marmots are so thin that their skin hangs loosely on their bodies. At this time they frantically feed on the stored hay and also the first fresh grass shoots so that they can regain their strength as quickly as possible for the breeding season ahead.

The observation that marmots apparently suffer no ill-effects from their hibernation in cold and damp conditions led to the false belief that marmot fat was an effective remedy for rheumatism. Partly for this reason they were killed in large numbers and this led to their extinction in many parts of their former range.

The related **Bobak Marmot** (*Marmota bobak*) lives in the steppe region of eastern Europe to west China and Siberia. It is somewhat larger than the Alpine Marmot. Like the sousliks, Bobak Marmots have been exterminated in many areas because of the damage they can do to crops. On the other hand they also help to aerate and mix the hard steppe soils and thus encourage plant growth.

Alpine Marmot gathering hay to line the nest

Friendly greeting between adult and young (above)
Marmots fight by pummelling with their front feet and baring their teeth (below)

European Souslik

Citellus citellus

(Squirrel Family)

Identification: HB 18-23 cm, T 5-7 cm, W 200-350 g. Male larger and heavier than female; fur yellowish or reddish-grey, often with black speckles on the back; tail short, somewhat bushy; ears small, scarcely appearing through the fur; emits a shrill whistle when alarmed.

Distribution: Southeast Europe, from Austria to the Ukraine.

Habitat: Dry steppe country, meadows and fields.

Behaviour: Diurnal; often stands bolt upright; digs deep burrows with branching tunnel systems; colonial; hibernates from Sep-Mar/Apr.

Food: Grasses, herbs, seeds, buds, fruits, insects, nestlings and eggs.

Breeding: Mating season Apr-May; gestation 25 days; 5-8 naked, blind young which open their eyes at about 8 days; young suckled for 3 weeks and independent at 2 months, but not sexually mature till the following spring.

The rat-sized European Souslik looks a bit like a small, slim marmot. It inhabits warm, dry steppe country, where is digs deep tunnels in the soil. Sousliks often live in large colonies. Each animal digs its own tunnel which it gradually enlarges until it forms a branching system of burrows up to 2 m deep. Like a hamster burrow this has a sloping entrance and a vertical emergency exit. In summer the young are looked after in a large lined chamber. Sousliks retire into their burrows in late autumn and hibernate after blocking the tunnels with hay and soil. They do not lay up winter stores, but survive entirely on their fat reserves for the five to seven months of hibernation.

European Sousliks have many enemies, of which the most prominent is probably the Steppe Polecat (p. 128), which specialises on this species in some areas. When an individual Souslik senses danger it makes a shrill warning whistle, at which sound all members of the colony rapidly make for cover. Other species also recognise this alarm call, including the Common Hamster. Sousliks were at one time blamed for spreading the plague, via fleas. This resulted in a major campaign to exterminate them, which succeeded only in spreading them further, along with the illness.

A second species, the **Spotted Souslik** (*Citellus suslicus*) lives further north and east, from southern Poland about as far as the Volga river. It has creamy-white spots on its back and a slightly shorter tail.

Spotted Souslik

Also on the dry Steppes of southeast Europe are several species of blind mole-rats including the **Greater Mole-rat** (*Spalax microphthalmus*) and the **Lesser Mole-rat** (*Nannospalax leucodon*). They are more closely related to rats and mice than to the sousliks. They have neither eyes nor external ears and live entirely underground. They feed on roots and bulbs and by pulling plants down by the roots into their tunnels.

European Souslik on the look-out (above)
Young European Sousliks at entrance to their burrow (below)

Common Hamster

Cricetus cricetus

(Mouse Family)

Identification: HB 22-30 cm, T 3-6 cm, W 150-400 g. Fur brown, white and black; black belly; white patches on face and shoulders; occasional all-black specimens or albinos occur; ears round, lacking hairs; threatens by snarling and chattering teeth.

Distribution: Central and eastern Europe; central Asia.

Habitat: Steppe, rough grassland.

Behaviour: Crepuscular and nocturnal; solitary; lives in branching underground tunnel system; carries food in cheek pouches; lays up stores for winter; hibernates Oct-Mar/Apr.

Food: Grasses and herbs, roots, fruit, grain; also insects, snails, slugs, worms, frogs, mice and young birds.

Breeding: Mating season Apr-July; gestation 19-20 days; 2-3 litters per year, each with 4-8 (rarely 12) naked, blind young. Young suckled for the first 18 days and independent at 4 weeks.

The Common Hamster is a familiar animal, although it is rarely seen in the wild. Originally it inhabited only the eastern European steppes, but now spread into cultivated parts of central Europe. The Common Hamster does not rely entirely on grain seeds but is a true omnivore. Transportable food is stuffed into the cheek pouches and then squeezed out again using the front feet, once back in the burrow. Over the summer the Common Hamster gathers enormous quantities (up to 15 kg) of stored food. The tunnel system consists of a metre-long entrance and one or more steep emergency burrows into which the hamsters can disappear quickly if danger threatens. These lead to a central nesting chamber lined with hay, usually 1-2 m deep. In late autumn hamsters seal the entrances with soil and begin their hibernation. At this stage their body temperature drops to about 5 °C. Every five or six days the hamster wakes up, nibbles at the stored food and uses a special toilet in the burrow. In the spring when the food reserves have been used up it re-emerges from the burrow. When

threatened Common Hamsters inflate their cheek pouches so the head seems three times as wide as normal. At the same time it makes a growling noise and chatters its teeth. When it sits bolt upright it displays its black belly. This reverse colouration may offer protection from enemies, since in most mammals the underside is much paler than the back.

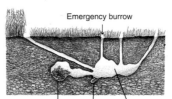

Emergency burrow

Foodstore   Latrine   Nest chamber

Cross-section of burrow system

Two other, smaller hamsters occur in southeast Europe: the **Rumanian Hamster** (*Mesocricetus newtoni*) is very similar to the familiar Golden Hamster which originated in Syria; the **Grey Hamster** (*Cricetulus migratorius*) is vole-sized and unpatterned.

Common Hamsters often sit bolt upright (above)
Common Hamsters can swim but rarely enter the water willingly (below)

Norway Lemming

Lemmus lemmus

(Mouse Family)

Identification: HB 13-15 cm, T 1.5-2 cm, W 35-100g. Colourful, with red-brown, yellowish and black patterning; considerable individual variation; ears and tail very short.

Distribution: Scandinavia.

Habitat: Tundra and Alpine zone above the tree line; birch and pine woods.

Behaviour: Mainly nocturnal; usually solitary; creates extensive above and below ground tunnel systems; runs very quickly and swims well; does not hibernate.

Food: Grass, sedge, berries, bark, roots, leaves.

Breeding: Mating season Apr-Sep; gestation 16-21 days; 2-3 (sometimes more) litters a year, each with 3-12 naked, blind young which are independent at 7 weeks.

Note: Every 3-4 years the population reaches a peak followed by a drastic reduction. In some peak years very high densities are reached and mass migrations occur.

Lemmings are well adapted to life in the Arctic tundra. Their rounded bodies and short legs and ears mean that they are well protected from the cold. They live amongst the ground vegetation in the tundra and also the subAlpine zone in the mountains, where they feed on roots and shoots. They tend to inhabit the tundra during the summer and migrate to the mountains during the winter, even in normal years.

At intervals, so-called 'lemming years', there are population explosions. Under these conditions, the young lemmings migrate in large numbers and become very bold and aggressive. At first they move only at night but soon they begin to travel by day as well, stopping at no obstacle, crossing streets, dams and rivers. Their numbers build up at lake shores and at the sea until pressure of numbers forces the leading lemmings to swim. Although Norway Lemmings swim quite well, the wind and waves drown them in their thousands. When they meet insurmountable obstacles, instead of returning the way they came they try to continue their journey using up all their energy until they fall down and die. These mass migrations therefore do not usually result in a permanent extension of the lemmings' distribution, although temporary colonies may become established beyond the normal range.

In peak lemming years all their predators do well, e.g. skuas, owls, Raven, Arctic Fox and Stoat. These species produce more young than usual which in turn continue to prey on the lemmings and contribute to the drastic crash in their population.

The somewhat smaller **Wood Lemming** (*Myopus schisticolor*) lives in the mossy coniferous forests of northern Europe. It has uniform slate-grey fur with a rusty-brown patch on the back. Unlike the Norway Lemming, this species rarely migrates. The Wood Lemming is unusual amongst mammals in feeding extensively upon moss.

The coat of a Norway Lemming is boldly patterned (above)
Young Norway Lemmings (below)

Bank Vole

Clethrionomys glareolus

(Mouse Family)

Identification: HB 8-12 cm, T 3.5-7 cm, W 15-40 g. Back chestnut brown; underside grey; ears broad and rounded, clearly projecting beyond fur; tail about half the length of the head and body.

Distribution: Whole of Europe except most of the Iberian Peninsula and northern Scandinavia. Absent from Iceland and much of Ireland, but widespread in Britain. Also found in northern Asia to central Siberia.

Habitat: Deciduous and mixed woodland; hedgerows; parks.

Behaviour: Mainly crepuscular; relatively tame; climbs well; develops system of burrows just beneath the surface of the ground or in the above-ground vegetation; constructs a circular nest at ground level.

Food: Grasses, herbs, fruits, seeds, bark; occasionally insects, worms.

Breeding: Mating season Mar-Sep, also during the winter in mild weather; gestation 18 days; 3-4 litters a year, each with 3-7 naked blind young which leave the nest at 3 weeks; young may reach sexual maturity at 9 weeks.

Bank Voles prefer a rich herb and shrub layer. They tend therefore to avoid bare, uncoppiced woodland and also cultivated areas. A Bank Vole will spend a lot more time at the soil surface than other species of vole and even builds its nest above ground if the soil is too loose or too hard. The nest is a sphere woven from grasses and other stems and lined with leaves and moss; it is usually anchored between thin twigs.

Bank Vole nest

This species also takes different food from the other voles, regularly eating insects and worms and sometimes eggs and nestlings or even fresh corpses.

Voles neither hibernate nor reduce their activity in the winter. In winter they construct special well-insulated nests and tend to restrict their foraging excursions to the early evening. At times of food shortage Bank Voles (and also Field Voles, p. 72) sometimes supplement their diet by stripping the bark from trees. Because of this they are sometimes regarded as pests by foresters.

There are two other similar voles occuring in Europe. The **Grey-sided Vole** (*Clethrionomys rufocanus*) lives in the mountains of Scandinavia and extends to the Far East. The slightly paler **Ruddy Vole** (*Clethrionomys rutilus*) replaces the Bank Vole on the tundra of north Scandinavia and has a circumpolar distribution including the tundra of North America.

Bank Voles are fairly easy to observe (above)
Grey-sided Vole: this species eats mainly buds and young shoots (below)

Common Vole

Microtus arvalis

(Mouse Family)

Identification: HB 9-12.5 cm, T 3-4.5 cm, W 14-50 g. Brownish or yellowish-grey above, pale grey underneath; tail about a third as long as the body; ears rounded and hairy towards the edge on the inside; call a high-pitched squeak.

Distribution: Southern and central Europe, but absent from Scandinavia and the British Isles except for particular sub-species on Orkney and Guernsey in the Channel Islands. Range extends east to Siberia.

Habitat: Open, dry fields and meadows, avoiding closed woodland.

Behaviour: Active by night and day; builds a tunnel system with many exit holes.

Food: Primarily grasses, supplemented by herbs, fruits, seeds, grain.

Breeding: Mating season Mar-Oct, throughout the year in some warmer parts of the range; gestation 19-21 days; 3-6 litters a year, each with 4-7 naked, blind young which are independent at 20 days old.

Note: Undergoes population explosions at 3-5 year intervals in parts of the range; in some areas a serious pest.

The Common Vole holds the record for speed of mammalian reproduction. After a three week gestation, the female produces six or seven young and often becomes pregnant again within a few hours of giving birth. The embryos of the next litter continue to develop even while the mother is still suckling the first brood. This means that, under suitable conditions, a female can produce a new litter every 20 days.

The young females can be mated when they are about 13 days old even though they continue to feed from their own mother until they are 20 days old. This situation is unique amongst mammals. In this way a single Common Vole can produce up to 500 young in its short life of just four or five months.

Even though Common Voles have many predators, including weasels, stoats, foxes, buzzards, kestrels and owls, this rate of reproduction can lead to population explosions, particularly when the weather is good and the food supply plentiful. In some areas these explosions occur on a regular cycle of three to five years. The females often gather together and bring up their young communally. Living in such close proximity, particularly in times of food shortage towards the end of the season, leads to a marked increase in aggressive behaviour. It can even result in mass starvation and cannibalism.

The voles on Orkney were probably introduced inadvertently by the first settlers about 5000 years ago, as shown by remains in Neolithic tombs. They are larger than anywhere else in Europe, perhaps due to the absence of ground predators such a weasels.

Tracks: hind paw (left) and front paw (right)

Common Vole, probably Europe's commonest mammal (above)
New-born Common Voles in the nest (below)

Field Vole

Microtus agrestis

(Mouse Family)

Identification: HB 9-13 cm, T 2.5-4.5 cm, W 18-60 g. Easily confused with Common Vole (p. 70). Dark or greyish-brown above with paler belly; very vocal, producing loud chattering or chirping calls.

Distribution: Central and northern Europe except Iceland and Ireland; eastwards to central Siberia. Widespread in Britain, including many Hebridean islands but not Orkney or Shetland.

Habitat: Damp, cool sites, woodland edges, parks, mountain meadows up to 2000 m, wet grassland, river banks.

Behaviour: Active by night and day; makes tunnels just underneath the surface of the soil and through thick grass; nests above ground in damp habitats.

Food: Grasses, herbs, tree bark in winter.

Breeding: Mating season Mar-Oct; gestation 21 days; 3-6 litters a year, each with 3-7 young.

Note: Undergoes population explosions like Common Vole, usually in a 3-5 year cycle. Sometimes damages trees by eating the bark.

Like all voles, the Field Vole has many enemies and is therefore always on its guard. It will often stop to sniff the air or to survey the scene by standing upright on its hind legs. It has acute hearing, but does not react to all sounds. For instance, the Field Vole easily becomes accustomed to traffic noise, but disappears instantly if it hears a twig snap or the rustle of leaves.

The position of the eyes at each side of the head give voles a wide visual field without needing to move their heads. Their eyes are also set quite high on the head so they can also scan the skies with ease and thereby see dangerous birds. Having said that, their eyesight is not that good and they rely more on an acute sensitivity to movement and vibrations.

Although Field Voles are often active by day they are much more difficult to observe than Bank Voles (p. 68). This is because they rarely leave the safety of cover provided by tall grass stems or a thick herb layer. However, their runs are so close to the surface of the soil that it is quite often possible to see them moving along beneath it. They also develop a system of regular tunnels and runs in amongst the thick vegetation. At intervals along these runs there are piles of olive green droppings and chopped sections of fresh grass blades.

Unlike the Common Vole, the Field Vole often makes its nest above ground, particularly in very damp soil. A Field Vole will defend their own territory fiercely against members of their own species. It is not unknown for the resident vole to bite or even kill the interloper.

During the breeding season male Field Voles produce a strong, musky smell which most people find unpleasant.

The Field Vole resembles the Common Vole (p. 70), with which it is often confused

Snow Vole

Microtus nivalis

(Mouse Family)

Identification: HB 10-14 cm, T 5-7.5 cm, W 40-65 g. Fur silver-grey; long whiskers; tail of adults whitish and held horizontally when running.

Distribution: High mountains of central and southern Europe; east to Iran.

Habitat: Rocky slopes to 4700 m.

Behaviour: Active by night and day; nests in clefts in the rocks, or digs short tunnels close to the surface of the ground.

Food: Strictly vegetarian, e.g. grasses and fruits.

Breeding: Mating season June-Aug; gestation 21 days; 1-2 litters a year; 2-4 young, independent at 3 weeks.

This rather tame species is relatively easy to see as it searches for food in its rather open habitat. Occasionally sunbathes on warm rocks. The Snow Vole's exceptionally long whiskers (up to 6 cm) help it negotiate narrow clefts in the rocks.

Energy balance is all-important in the survival of small animals at high altitudes. In cold, damp weather Snow Voles stay in their burrows and thereby avoid rapid temperature loss. They often shelter in mountain huts. They do not hibernate, but often stay under snow cover where they are protected from hard frosts and storms.

Common Pine Vole

Microtus subterraneus

(Mouse Family)

Identification: HB 8-10 cm, T 2.5-4 cm, W 13-23 g. Fur slate-grey to greyish-brown, somewhat paler below; eyes very small; ears almost completely covered by fur.

Distribution: In France through southern Germany to the Balkans. Not in Britain.

Habitat: Damp sites, meadows, deciduous woodland; gardens; mountains to 2000 m.

Behaviour: Mainly nocturnal; digs shallow tunnel system and runs in thick vegetation; nest usually below ground.

Food: Leaves and flowers, seeds, berries, fungi.

Breeding: Mating season Apr-Oct, sometimes throughout the year; gestation 21 days; 2-9 litters a year with 2-3 young, which are mature at 8-9 weeks.

This species lives in a wide variety of habitats, but usually only occurs where the Common Vole (p. 70) is absent. Possibly it is unable to compete with the higher reproductive rate of the Common Vole. Its average of two or three nestlings per litter is unusually low for a vole.

Although the Common Pine Vole is quite widespread, it is rarely seen in the wild. This is because it hardly ever leaves cover and usually takes its food back to the nest before eating it. In addition to digging its own burrows, this species often uses old Water Vole (p. 76) tunnels. In heavy rain or snowy conditions it seals the burrows from the inside with soil.

Several closely related species occur in southern Europe in Iberia, southern France, Italy and the Balkans, some occupying very small ranges.

Snow Vole (above)
Common Pine Vole (below)

Northern Water Vole

Arvicola terrestris

(Mouse Family)

Identification: HB 12-20 cm, T 7-11 cm, W 80-180 g. Colour very variable, ranging from yellowish-brown to almost black; tail more than half the length of the body; ears covered by fur.

Distribution: Europe except Iceland, Ireland, much of France and most of Spain and Portugal; Asia. Sparsely distributed throughout Britain.

Habitat: Banks of slow rivers or lakes, in some parts of range pastures, fields and even gardens.

Behaviour: Active by day and night; swims and dives well; digs an extensive burrow system with living quarters and nest.

Food: Bank and water plants, roots, fruit.

Breeding: Mating season Mar-Oct; gestation 21 days; 3-4 litters a year, each with 2-10 young which are independent at 4 weeks old.

The Northern Water Vole is the largest native European vole. Only the introduced Muskrat (p. 78) is bigger. Often confused with Common Rats, especially the darker coloured forms. The water vole is distinguished from the rat by its shorter nose and much shorter tail.

The much-branched tunnel systems are often so close to the surface that the vole's movements can be followed. Occasionally the vole will push soil to the surface in mounds rather like a mole. It also uses mole tunnels in order to reach plant roots without having to dig. Their liking for roots and buds means that they are often regarded as a pest by gardeners. Such plant material is difficult to digest and must first be broken up into very small pieces. The molar teeth of voles are particularly well adapted to this task. The teeth of each species are in a particular pattern which is useful in identification.

Water Voles favour habitats with a lot of thick vegetation. They can both swim and dive very well. In addition, they can close their mouths behind their incisor teeth and are thus able to chew plants while submerged under the water.

The very similar but somewhat larger **Southern Water Vole** (*Arvicola sapidus*) is found on the Iberian Peninsula and also in western France.

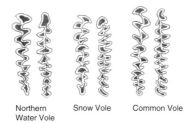

Northern   Snow Vole   Common Vole
Water Vole

The particular pattern of molar teeth is characteristic
for each species of vole. Upper teeth are on
the left of each pair, lower teeth on the right.

The Northern Water Vole is almost rat-sized (above)
They prefer to live close to water (below)

Muskrat

Ondatra zibethicus

(Mouse Family)

Identification: HB 25-40 cm, T 20-25 cm, W 600-1800 g. Fur dark brown, slightly paler on the belly; ears round, scarcely visible above fur; tail virtually hairless and laterally compressed; smells distinctly musky.
Distribution: Central and eastern Europe (introduced from North America); also found in Finland and adjacent parts of Russia.
Habitat: Banks of still bodies of water with thick vegetation.
Behaviour: Mainly active at dusk and by night; swims and dives well; constructs reed mounds and tunnels; lives in loose colonies.
Food: Aquatic and marsh plants; molluscs, crustaceans.
Breeding: Mating season Apr-Oct; gestation 28 days; several litters a year, each with 5-8 naked, blind young; young suckled for 3 weeks and mature at 3-5 months.

The Muskrat is not native to Europe but originates in North America where it has long been hunted for its pelt (known as musquash), which is unusually thick and soft. Since early this century Muskrats have been farmed in Europe for their fur. Escapes have enabled wild populations to become established in many parts of Europe. Colonies in Britain and Ireland became established in the 1930s but were eliminated within a few years by an intensive trapping campaign.

The entrances to Muskrat tunnels normally lie below the water surface. This is also true for the entrances of the conical mounds of plant material that they build. These mounds can be up to a metre high and are frequented particularly in the winter. It is here that Muskrats undergo a short hibernation when weather is severe and also store their food.

Muskrats can cause considerable damage to dams and flood barriers, and for this reason there have been various (mainly unsuccessful) attempts to exterminate them. They can build up their populations quickly with their high reproductive rate. They have also been helped by the decline of their natural enemies such as otters, mink, fox and sea eagle. Indeed there are probably more Muskrats now living between France and China than in the whole of the USA and Canada. In the former Soviet Union and in Finland Muskrat colonies are sometimes specifically protected for the benefit of the fur trade.

Muskrats are particularly well-adapted to their aquatic life. Their hind feet are partially webbed and fringed with bristles, making excellent paddles. Only the hind legs are used for swimming; the front limbs are held close to the body, while the tail wags backwards and forwards in the water. When a Muskrat dives, its ears are closed by a fold of skin and the mouth can also be shut behind the incisor teeth. It can therefore continue to feed whilst holding its breath beneath the water. Muskrats take their name from the strong musky secretion with which the male marks out his territory.

Muskrat: note shaggy appearance of fur (above)
Winter lodge of reeds (below)

European Beaver
Castor fiber

(Beaver Family)

Identification: HB 75-100 cm, T 30-35 cm, W 18-30 kg. Largest European rodent; fur very dense; grey to blackish-brown; tail broad and flattened horizontally, scaly; ears and eyes small; hind feet webbed.

Distribution: Earlier found in most of Europe and central Asia, exterminated in Britain by the 13th century. Today restricted to a few areas: Rhone, Elbe, southern Norway, Finland, Poland; reintroduced to a number of sites, in France, Switzerland and southern Germany for example.

Habitat: Still and slow-flowing water with adjacent woodland.

Behaviour: Crepuscular and nocturnal; European Beavers will pair for life and live in family groups; they are excellent swimmers and can dive for up to about 15 minutes. They nest in burrows in the river bank or sometimes build elaborate lodges using branches, twigs and mud, with entrances always beneath the water. They are capable of felling trees of up to 80 cm diameter and regulate the water level by building dams. They do not hibernate.

Food: Entirely vegetarian, feeding on grasses, leaves, twigs and bark; also soft wood.

Breeding: Mating season Jan-Feb; gestation 15 weeks; 1-5 fully furred, sighted young, capable of swimming and diving within a few days of birth; young suckled for 2 months and remain in family group for 2-3 years.

Beavers are well-known for their tree-felling and dam-building activities. Nevertheless, the details of their biology have only recently come to light and there are many incorrect rumours and stories about them.

Beavers have long been hunted not only for their valuable fur but also for the secretions of their musk glands. This secretion (castorium) was formerly used in traditional medicine and in the manufacture of perfumes. Both male and female beavers possess a pair of these large musk glands beneath the tail. The yellow, oily secretion has an intense but not unpleasant smell. Beavers use the secretions to mark out their territories. Each Beaver family consists of a pair of adults which remain together throughout their lives, plus their offspring of varying age. Young leave their families at two or three years old (sometimes as late as four years) in order to found a family of their own.

A Beaver's lodge, which contains the living quarters, is usually large and stable enough to carry the weight of a person. Several tunnels lead out from the living quarters down into the water. (Continued p. 82.)

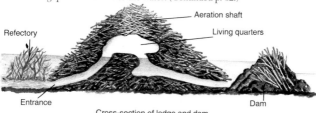

Aeration shaft

Living quarters

Refectory

Entrance

Dam

Cross-section of lodge and dam

European Beaver at water's edge: note flattened, scaly tail (above)
Feeding time (below)

European Beaver (continued)

The branches and twigs that make up the lodge are carefully cemented together with mud and clay, except for a vertical chimney leading upwards that serves as a ventilation shaft. Beavers gather building materials by night on the river bank where they also cut down trees. With their chisel-like incisors they can gnaw through an 8 cm wide tree trunk in only five minutes. Larger trees often take days or even weeks, and several beavers work together at the task. They tend to prefer soft wood such as poplar, aspen, alder or willows; these are the species which tend also to be of less importance to people.

As is the case in all rodents, Beavers' incisor teeth continue to grow throughout their life and are only kept to their normal size by repeated use. The leaves and twigs of felled trees are consumed and the trunks and branches are cut up into shorter lengths before being transported to the lodge or dam. In the autumn, Beavers carry quite a few branches and anchor them in the mud at the bottom of a lake. The cold water then acts rather like a fridge and keeps them fresh during the winter. The animals usually spend very cold periods out of sight beneath the ice. In winter the water level often drops, leaving air pockets between the water surface and the ice. At this time of year Beavers mainly feed on the bark of stored twigs and branches which they shred inside the lodge. They can continue to gnaw whilst submerged without taking in water by sealing the back of the mouth and throat with folds of skin. During a dive they can also close their nostrils and ears and their eyes are covered by a transparent membrane.

Beavers swim using their large hind legs. The flat tail functions as a rudder. When danger threatens, they use their tails to make a loud splash on the water surface. Until the last century monks used to regard the scaly tale of the beaver as an indication of its close relationship to fish. Fish were allowed to be eaten during fast periods, and Beaver was a welcome addition to their diet.

In Finland and eastern Europe the related **Canadian Beaver** (*Castor canadensis*) has been introduced. It is difficult to distinguish from the European Beaver externally. This species breeds more rapidly than the native European Beaver and has also produced hybrids with that species.

Hind foot (left) showing webbing. Front foot (right)
showing opposable digit for gripping twigs

European Beaver swimming with freshly cut branch (above)
Lodge (centre)
Tree felled by beavers (below)

Coypu

Myocastor coypus

(Coypu Family)

Identification: HB 40-65 cm, T 30-40 cm, W 7-10 kg. Female somewhat smaller than male; fur dark brown to grey-brown; tip of nose white; tail rounded in cross-section, scaly, sparsely hairy; hind feet webbed; large orange incisors, which are often visible.

Distribution: Native to the southern half of South America; repeatedly introduced to central and western Europe for fur-farming. Feral in southeast England from the 1930s until exterminated in 1990 by an intensive trapping campaign.

Habitat: Rivers and streams with thick bankside vegetation.

Behaviour: Mainly active at dawn and dusk; usually lives in pairs, sometimes in small colonies in bankside holes which they dig out for themselves; swims and dives exceptionally well, but rather ungainly on land.

Food: Water plants and marsh plants.

Breeding: Able to mate at any time of the year; gestation 130 days; 2-3 litters a year, each with 4-6 (can be as many as 12) fully developed young which are independent at 3 months old.

Note: This species has been bred in various colour varieties in fur farms, including albino forms.

Coypus are not closely related either to Beavers or Muskrats: they are closer to Guinea Pigs. They were originally introduced to Europe for their pelts (known as Nutria) but a proportion of the animals escaped and set up colonies in the wild. In size Coypus are between Beavers and Muskrats and like them are restricted to the vicinity of water. At the slightest disturbance they take to the water and can remain submerged for up to five minutes at a time.

Coypus usually live in pairs and dig short tunnels in the bank with an enlarged living chamber at the end. If the bank is not suitable for tunnelling they can also make nests in the water using reeds (see also Muskrat, p. 78).

Young Coypus are much further developed at birth than the young of native European rodents other than the Beaver. They have their full complement of hair and can already see well. The teats of the mother are set high on the sides of the body so that she can suckle her young even when they are all swimming. Coypus constantly tend their fur and clean it to make sure it remains waterproof. They spread oil from special glands at the

Young feeding from female in the water

corners of the mouth using their paws and distribute this over the entire coat, combing it thoroughly with their paws. Coypus sometimes die in harsh weather if the water freezes for a long period.

Coypu at water's edge: note rounded tail (above)
Female with young: note bright orange canine teeth (below)

Fat Dormouse

Glis glis

(Dormouse Family)

Identification: HB 13-19 cm, T 11-15 cm, W 80-120 g, but almost twice this weight just before hibernation. Ash-grey to grey-brown above, white beneath; large dark eyes; ears fleshy and rounded; tail bushy; makes noisy squeaks during breeding season.

Distribution: Southern and central Europe; east to Iran. Introduced to England at Tring, Hertfordshire in 1902 where they are now well established but limited to an area of about 600 km^2.

Habitat: Deciduous and mixed woodland, parks, gardens.

Behaviour: Active at dawn and dusk and during the night; sleeps by day in a leafy nest in hollow tree or nest box; climbs well amongst branches and usually stays in the trees; long hibernation Oct-Apr/May, usually in hole in the ground.

Food: Shoots, leaves, bark, berries, nuts, acorns, beechmast and other seeds.

Breeding: Mating season June-Sep; gestation 30 days; 2-10 (usually 4-6) naked, blind young, suckled for 3 weeks, after which they open their eyes.

Fat Dormice can sometimes be heard pattering about in the attics of houses close to woodland as they begin their night-time foraging. They clamber amongst the tree branches like small goblins and their large dark eyes mark them out clearly as nocturnal animals.

In the autumn they eat large quantities of seeds, nuts and fruit and build up a thick layer of fat, doubling their normal body weight. After the first frosts they dig out special winter quarters which are up to a metre deep in the soil. Sometimes they hibernate in barns or lofts, more rarely in nest boxes or hollow trees. During hibernation they roll up with the tail over the head and become cold and stiff. Their body temperature drops to as low as 1 °C and pulse and breathing are scarcely detectable. In this state the body is just ticking over and needs very little energy except that provided by the gradual using up of the fat reserves. Sometimes Fat Dormice will hibernate for up to seven months of the year. When they finally wake in May or June they have lost up to fifty per cent of their body weight.

The Romans regarded the Fat Dormouse as a delicacy and fattened them up for special meals, hence the alternative name Edible Dormouse.

Fat Dormouse in hibernation

Female with new-born young in leaf nest (above)
When disturbed the mother will carry her young individually to safety (below left)
Young Fat Dormice emerging from nest hole (below right)

Garden Dormouse

Eliomys quercinus

(Dormouse Family)

Identification: HB 11-17 cm, T 9-13 cm, W 50-120 g, in autumn to 180 g. Fur reddish-brown with white on the underside and black markings on the face and head; tail long and somewhat bushy, dark above and with white tuft at tip. Ears large and leathery.

Distribution: Southern and central Europe, parts of eastern Europe, related species in southwest Asia.

Habitat: Deciduous, mixed and coniferous forests, scrub, vineyards; in mountains to 2200 m.

Behaviour: Nocturnal; climbs well and also found on the ground. Builds nests in tree holes, rock crevices or nest boxes and usually digs a hibernation hole in the ground.

Food: Almost entirely predatory: butterflies and moths, grubs, beetles, crickets, grasshoppers, spiders, slugs, small birds, eggs, nestlings and young mice. In the autumn mainly fruits, seeds and berries.

Breeding: Mating season Apr-Sep; gestation 23 days; 3-7 naked, blind young which open their eyes at 18 days old; young suckled for 4 weeks and leave the nest at 30 days.

The Garden Dormouse builds its nest either in the ground or in a tree. Unlike that of the Fat Dormouse it is not lined with leaves but with grass and moss. Sometimes several pairs of Garden Dormice nest close together in a loose colony. Females use scent markings to indicate their individual territories which they defend vigorously.

The young are born in a relatively helpless state after a short period of gestation and they do not open their eyes until they are 18 days old. They lick saliva from their mother's mouth and this probably provides them with important nutrients, as does the mother's milk. The young grow very quickly and soon leave the nest. Sometimes the female leads the young in a caravan formation (see also some shrews, above). The young are independent in four to six weeks, after which they must feed quickly in order to build up sufficient reserves to see them through their first hibernation.

Like all dormice Garden Dormice do not lay down winter stores. In the autumn they switch their diet from small animals and soft fruits to nuts, acorns, beechmast, chestnuts and the like. When the frosts begin several animals usually gather together in a winter nest for hibernation. The site may be a crevice in the ground, the tunnel of some other rodent, or a hollow tree or cave.

The female sometimes leads her
young in a caravan formation

Garden Dormouse in typical orchard habitat

Forest Dormouse
Dryomys nitedula

(Dormouse Family)

Identification: HB 8-11 cm, T 8-9 cm, W 17-40 g. Smaller than Fat and Garden Dormice; fur grey or brown, pale beneath; black face mask; tail uniform in colour and bushier than that of the Garden Dormouse; ears short and rounded.

Distribution: Central, eastern and southern Europe.

Habitat: Deciduous, mixed and coniferous woods, in mountains to 3500 m.

Behaviour: Nocturnal; builds rounded nests from grass and twigs, usually about 1-2 m high in a bush; also nests in hollow trees or old bird or squirrel nests; climbs well; digs hole in the ground for hibernation.

Food: In summer mainly predatory: moths, grubs, beetles, birds' eggs. In autumn mainly acorns, beechmast, nuts, fruit and berries.

Breeding: Mating season Apr-Aug; gestation 23-25 days; 2-6 naked, blind young which are independent at 4-5 weeks old.

Note: In warm areas this species does not hibernate and produces 2-3 litters a year.

The Forest Dormouse uses its sharp claws to cling on to twigs and branches as it climbs through the undergrowth. As another adaptation to climbing, the bare soles of its feet have tiny, soft ridges which can get a grip on even quite smooth surfaces and prevent slipping. Nevertheless, the smooth bark of beech trees sometimes defeats them and they have to go up another tree in order to climb up into the branches. The long tail acts as a balancing counterweight when climbing thin twigs or jumping. Forest Dormice lose their tail skin, and sometimes a few vertebrae as well, very easily – a useful anti-predator device shared with the other dormice.

Like other dormice the Forest Dormouse switches its diet in the autumn in order to lay down fat reserves for hibernation. The hibernation site is usually a hole deep in the ground or a hollow tree. Active throughout the year in warmer regions.

In the Alps the Forest Dormouse is slate-grey in colour, as it is in the Balkans and Carpathians to central Asia. Further north and east the normal colour is grey to reddish-brown.

Forest Dormouse (left) and Garden Dormouse
(right). Note shorter ears and smaller face
mask of Forest Dormouse.

Forest Dormice can clamber on the thinnest branches and twigs
because they weigh so little: note uniform grey bushy tail

Common Dormouse

Muscardinus avellanarius

(Dormouse Family)

Identification: HB 7-9 cm, T 7-8 cm, W 15-40 g. Fur yellowish-brown; eyes large and dark; ears short and rounded; tail with thick covering of hairs.

Distribution: Mainly southern and central Europe, eastwards to central Russia; in the British Isles only in southern England, southern Wales and Cumbria.

Habitat: Deciduous and mixed woodland with a good rich undergrowth; woodland edges, especially those with berry-bearing shrubs.

Behaviour: Nocturnal; climbs, jumps and runs with great agility; fairly solitary when adult. Constructs spherical nests from grass and leaves, usually sited in a shrub or low tree, sometimes in a nest box or hollow tree; usually hibernates below ground.

Food: Buds, shoots, berries, fruits, seeds; also insects.

Breeding: Mating season Apr-Aug; gestation 23 days; 1-2 litters a year, each with 2-9 (average 3-5) naked, blind young which open their eyes at 18 days old and are independent at 5-6 weeks.

Common Dormice are sometimes mistaken for young squirrels, but they are not closely related. This is our smallest dormouse and, like the others, it has a furry tail. The tail is normally used as a counterbalance, but is sometimes also helpful as a strut, almost like an extra limb.

The characteristic grass and leaf nest can often be found in hazel scrub or in blackberry or raspberry bushes, usually 1-2 m above the ground. Each Common Dormouse usually builds one or more of these summer nests and uses them in rotation. If disturbed, a dormouse will leave the nest immediately, but usually remains in the area, crouching motionless on a branch before creeping away. Nests in which the young are reared are usually about twice as big, that is about 12 cm across.

Spherical nest of a Common Dormouse

Like other dormice, the Common Dormouse hibernates in the winter. In October the well-fed dormice retreat to their winter nests which are usually in or on the soil or amongst roots. Sometimes several animals hibernate together. If the winter is very fierce such that even the soil freezes, many Common Dormice do not survive.

Dormice use their large eyes to help them see during the night; their bodies are also well supplied with long, sensitive bristles (above)
Dormice sometimes nest in holes in trees (below)

House Mouse

Mus musculus

(Mouse Family)

Identification: HB 7-10 cm, T 6-10 cm, W 10-30 g. Fur greyish-brown, slightly paler below; tail almost hairless and about the same length as the body; ears large, rounded and fleshy; unpleasant sharp smell.

Distribution: Originally Asia, but now found worldwide, usually in association with people.

Habitat: Mostly in urban habitats, houses, shops, factories, warehouses and the like; also sheds, cellars, open fields and hedgerows.

Behaviour: Active mainly by night and sometimes also by day; climbs and swims well; lives in family groups; in buildings makes nest from chewed-up small pieces of paper, material or the like; in more natural habitats digs its own burrow in the soil.

Food: Omnivorous; prefers seeds and grain.

Breeding: Mating season throughout the year, in more natural habitats Apr-Oct; gestation 20-24 days, 4-8 litters a year, each with 4-9 naked, blind young; young open their eyes after 13 days; are suckled for about 3 weeks and are mature at 6-7 weeks.

Note: There are many domesticated colour varieties, including albino.

There are two distinct sub-species of House Mouse in Europe: a western, greyer form usually found in buildings, and an eastern, brown-backed form which tends to live in meadows and fields during the summer, moving to houses and buildings during the winter. Whereas wild House Mice often store food in their burrows, those living in houses seem mostly to have lost this habit; instead they rely on human stores of food.

Mouse communities feature a definite dominance hierarchy. Each family is led by a strong, dominant male who alone pairs with the females. This dominant male maintains his position by fierce disputes with other males until he is eventually overpowered and superseded by another, usually younger, male.

As with rats, female House Mice often have communal nests with mixed groups of young. This helps the young mice to maintain their body temperature and survive cool conditions. They need a nest temperature of about 30 °C.

Heads of Common Vole (left) and House Mouse

House Mice have a special mechanism for regulating their populations. At high density, the female mice lower down the hierarchy lose their fertility and the sex organs of the developing young females also fail to mature properly. This occurs in other rodents, but most intensely in House Mice. House Mice can be a serious pest, particularly of stored grain and other crops. They are less important as vectors of human diseases, but can carry salmonella and also transmit Weil's disease.

Other, closely related species of mice, with white bellies and usually shorter tails occur in outdoor habitats in Iberia and southeast Europe.

House Mouse; note relatively hairless tail, about the same length as body (above)
Nest with two litters of different ages (below)

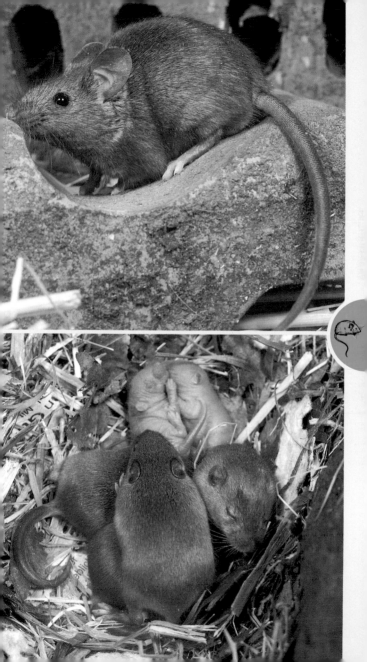

Common Rat
Brown Rat
Norway Rat

Rattus norvegicus

(Mouse Family)

Identification: HB 22-26 cm, T 18-22 cm, W 250-500 g. Larger and dumpier than the Black Rat (p. 98); fur grey-brown to almost black, pale grey underneath; feet flesh-coloured; tail virtually hairless, rounded in cross-section and scaly, always shorter than the body; ears rounded, fleshy.

Distribution: Originally eastern Asia, but now found all over the world and ubiquitous in Britain and Ireland.

Habitat: Towns and cities, cellars, sewers, rubbish dumps; often found close to water, in ditches, on river banks and at the sea shore.

Behaviour: Mainly active at dawn, dusk and by night; climbs, swims and dives well; lives in family groups with a strict dominance hierarchy; nests in buildings or digs its own burrow system in the soil.

Food: Omnivorous: seeds, green plants, fruit, human food, mice, birds, eggs, fish and carrion.

Breeding: Mating season throughout the year, in colder regions Apr-Oct; gestation 22-24 days; 4-6 litters a year, each with 6-12 naked, blind young; young open their eyes at 15 days, are suckled for about 3 weeks and are independent at 6-7 weeks.

Note: The albino form of this species is used as a laboratory animal.

The Common Rat arrived in Europe in the early part of 18th century – the first English record was in 1728. It was introduced to North America by 1775 and transported via ships throughout the world.

It usually lives in groups of up to 60 related animals, who can recognise each other by smell. Unrelated intruding rats are violently expelled from the group and territorial fights can result in fatal injuries.

The overall success of the Common Rat is a direct result both of its adaptations to a wide variety of habitats and its high reproductive rate. Young rats are sexually mature at three months. Within the groups the adults do not live in pairs, but the fertile females mate with many different males. Large litters are the norm, and up to 20 young have been recorded. More than one female

Common Rat (below) and similar young Muskrat (above)

may however use the same nest and they may bring up their young communally. Under these circumstances, if one of the mothers is killed or injured the young do not starve.

Rats can be dangerous to human health because they carry disease organisms (notably salmonella and Weil's disease), especially since they often live in rubbish tips.

Common Rats are adept climbers

Black Rat
Ship Rat

(Mouse Family)

Identification: HB 16-22 cm, T 18-24 cm, W 150-250 g. Similar to Common Rat (p. 96), but noticeably smaller and slimmer; coat black or grey-brown, grey or white underneath; feet flesh-coloured; tail almost hairless, scaly and always longer than the body; ears large, rounded and fleshy.

Distribution: Originally southern Asia, but now found throughout the world, especially in warm, temperate and tropical areas. In Europe common only in the south, sporadically in sea ports elsewhere. In Britain it has been replaced almost entirely by the Common Rat and survives only on a few small islands such as Lundy.

Habitat: Dry roof spaces, granaries, warehouses and boats; in southern areas also in open country such as orchards and scrub.

Behaviour: Active at dawn, dusk and by night; climbs very well; lives in large groups without strict hierarchy; constructs nest from waste material such as sawdust, paper or fragments of cloth; in wilder areas uses grass and leaves.

Food: Omnivore, but prefers plant material such as seeds and fruit.

Breeding: Mating season Mar-Oct; gestation 22-24 days; 2 litters a year, each with 6-12 naked, blind young; for development see Common Rat. (p. 96).

Like the Common Rat, the Black Rat is originally from Asia. It was however introduced much earlier to Europe (in Britain by the third century AD). This species is more dependent on warmth and is more closely associated with people, particularly in cooler regions. It was originally a tree-dwelling species and this may explain why it is more at home in lofts and attics. It is also known as the Roof Rat. The Common Rat by contrast is much more comfortable at ground level. Black Rats are mainly active at night. They leave tell-tale signs in the form of dark marks on balconies and floorboards. Such runways are a result of repeated marking with droppings and urine which are then further spread by the rats' feet. Almost all their movements are restricted to these runways.

The Black Rat is the species responsible for outbreaks of plague, especially during the Middle Ages. However, unlike the Common Rat, this species does not carry Weil's disease, although it can carry typhus. Under good conditions, and particularly in places where grain and other foods are stored, Black Rats can increase their populations very rapidly. Black Rats are also known as Ship Rats because of their association with boats, and the major ports and harbours of the world are still strongholds of this species. However, this species has become much less common inland in central and western Europe, perhaps because fewer houses are now made of wood and buildings are generally better sealed.

Black Rats are good climbers so ship mooring ropes are often fitted with conical structures to prevent rats from climbing them

When cornered the Black Rat can be quite ferocious

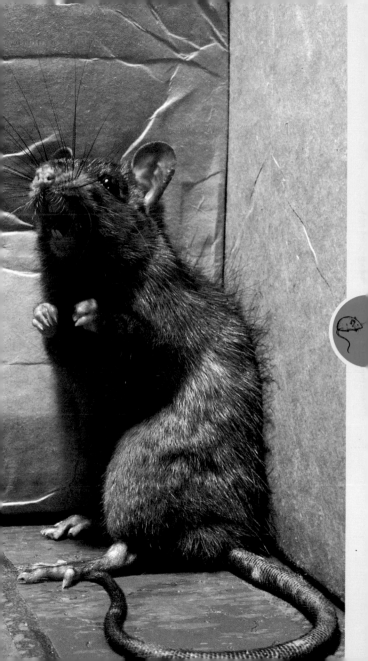

Harvest Mouse

Micromys minutus

(Mouse Family)

Identification: HB 5-7.5 cm, T 5-7.5 cm, W 4-10 g. Fur yellow to rusty brown, white beneath; tail almost hairless, about the same length as body, used as extra limb; ears small and rounded.

Distribution: Most of Europe except the far north and south, temperate parts of Asia. Absent from Iberia, most of Scandinavia, Ireland and most of Scotland but widespread in England and parts of Wales.

Habitat: Damp meadows with tall grass, well-vegetated river banks and lakesides, ditches, scrub, cereal crops; in winter often found in barns and storehouses. Absent from hills and mountains.

Behaviour: Active by day and night; climbs well up stalks and twigs and uses its tail as an extra support; constructs spherical grass nest in summer at a height of 20-100 cm. Nest anchored by being woven around vertical stems. Winter nest on or under the ground; sometimes in buildings.

Food: Seeds of grasses and other plants, grain, berries, buds, shoots and insects.

Breeding: Mating season Apr-Sep; gestation 21 days; 2-4 litters a year, each with 4-7 naked, blind young which open their eyes when 8-10 days old; young suckled for 16 days and mature at 5 weeks.

Although not rare over much of its range the Harvest Mouse, being very shy, is quite difficult to spot. Also, since it is active by day as well as by night it has many enemies. When in danger it is capable of freezing and remaining motionless for long periods amongst the stalks of grasses and tall herbs. In an extreme emergency a Harvest Mouse will fall vertically to the ground and make a quick getaway under cover.

The sleeping or breeding nest is woven into neighbouring grass stalks and consists of shredded grass, leaves and other plant material. The breeding chamber has a soft inner lining of finely chewed leaves. Young Harvest Mice are extremely tiny, weighing less than a gramme. At this stage they are blind, naked and completely helpless. However, they are independent by the time they are 16 days old, at which stage they leave the nest. The female builds a new nest for each litter.

Harvest Mice require warm conditions and like to sunbathe, either on top of the nest or on a leaf. Their territories are very small and normally cover just a few square metres.

When danger threatens, Harvest Mice remain completely motionless in a vertical position between the grass stems and in this state are easily overlooked

Harvest Mice use their tails to help them clamber amongst the stalks (above and centre)
Harvest Mice at their nest (below)

Wood Mouse

Apodemus sylvaticus

(Mouse Family)

Identification: HB 7.5-11 cm, T 7-11 cm, W 20-30 g. Fur yellowish-brown, pale grey beneath with at most a narrow streak of brown underneath; feet whitish; tail sparsely hairy and whitish on the underside; eyes and ears very large.

Distribution: The whole of Europe except northern Scandinavia; northwest Africa; southwest Asia. Widespread in Britain and Ireland.

Habitat: Fields, scrub, woodland edges and clearings, parks; sometimes in buildings especially during the winter; in mountains up to 1000 m.

Behaviour: Active at dawn and dusk and during the night; jumps and climbs well, but usually stays on the ground; digs deep burrows in the earth, with separate feeding chamber and nest; sometimes uses abandoned vole tunnels and occasionally nests in holes in trees; not very sociable.

Food: Seeds of grasses, herbs and trees, buds, shoots, berries, fungi and also insects, worms and slugs.

Breeding: Mating season Mar-Oct; gestation 23 days; 3-4 litters a year, each with 3-8 naked, blind young; young open their eyes after 13 days, are suckled for 2-3 weeks and are mature at 2 months.

Mice are distinguished from voles by their larger eyes and ears, more pointed nose and long, almost hairless tail.

As its name suggests, the Wood Mouse is found in woodland, but also lives in more open country such as arable fields, gardens, scrub and on sand dunes. The Wood Mouse often sits up on its haunches. When danger threatens Wood Mice can cover the ground remarkably quickly with long hops. Each jump may be up to 80 cm.

The somewhat larger and ash-grey **Rock Mouse** (*Apodemus mystacinus*) lives in the dry, rocky karst country of southeast Europe and Turkey. This species does not dig tunnels but builds its nest in rock crevices. It has very long whiskers (up to 5 cm) that help it to negotiate the narrow rock crevices, particularly at night. Other closely related species occur in the Alps and in eastern Europe.

Wood Mice often stretch up and stand on their toes in order to get a better view of their surroundings

When disturbed, the mother mouse moves her young to a safer site

Wood Mouse: note large eyes and ears, pointed snout and long tail

Yellow-Necked Mouse

Apodemus flavicollis

(Mouse Family)

Identification: HB 8.5-13 cm, T 10-13 cm, W 22-45 g. Fur yellowish-brown above and white underneath, with a sharp border between the two colours; usually shows a conspicuous broad yellow patch on the throat; eyes and ears very large.

Distribution: Central, eastern and southeast Europe; absent from much of Iberia and France and in Britain found only in Wales and southern England.

Habitat: Deciduous and mixed woodland, scrub, parks; found in houses in the winter; in mountains right up to tree line (2000 m).

Behaviour: Active at dawn, dusk and by night; jumps and climbs very well; not very sociable; nests amongst roots, in holes in trees or nest boxes and occasionally below the ground; lays down large winter stores.

Food: Seeds, fruits, fungi, shoots, buds, insects.

Breeding: Mating season Feb-Oct; gestation 23-25 days; 3-5 litters per year, each with 4-7 naked, blind young; young open their eyes at about 13 days, are suckled for 3 weeks and are mature at about 3 months old.

The Yellow-Necked Mouse is extremely similar to the Wood Mouse, but is usually somewhat larger. It is much more associated with trees or at least with woody shrubs. It often climbs amongst the branches and may site its nest in an old bird's nest or nest box. Other favourite nesting sites are between rocks and in amongst tree roots. The nest is lined with leaves and grass. This species often enters houses during the winter months, typically making its home in the roof space or top storey.

The Yellow-Necked mouse should never be picked up by the tail because it will immediately shed the skin and make good its escape. Other rodents that show this anti-predator tail-shedding behaviour include the Wood Mouse (p. 102), Striped Field Mouse (p. 106), Spiny Mouse (p. 106) and the dormice (from p. 86). The skinless tail rapidly dries up and falls off leaving a shorter stumpy tail behind.

Mice transport large numbers of tree seeds, such as various nuts and acorns, and in this way help in the distribution and regeneration of forest species

Yellow-Necked Mouse showing yellow collar

Striped Field Mouse

Apodemus agrarius

(Mouse Family)

Identification: HB 9.5-12 cm, T 7-8.5 cm, W 15-25 g. Fur yellowish or reddish-brown, paler underneath with a clear dark line along the centre of the back; tail almost hairless and shorter than body.

Distribution: Eastern and southeast Europe, northern Germany, Alps.

Habitat: Scrub, woodland margins, hedges, gardens, parks, river banks; in winter often in sheds and store houses; in mountains up to about 900 m.

Behaviour: Mainly active during the day; runs very quickly and climbs well; lives in small family groups; digs shallow tunnels with separate living quarters and nest.

Food: Omnivorous: seeds, berries, fruit, insects, worms, molluscs.

Breeding: Mating season Apr-Sep; gestation 21 days; 3-4 litters a year, usually with 5-8 young which are suckled for 2 weeks and independent at 3 weeks, becoming mature at about 8 weeks.

The most noteworthy feature of this species is the 4 mm wide black stripe running from the forehead right to the base of the tail. Like other mice it does not eat much in the way of green vegetable food such as grass stems but concentrates almost entirely on seeds and fruit. In spring it supplements this diet with small animals as well: mainly insects, molluscs and worms, even on occasion frogs.

Spiny Mouse

Acomys minous

(Mouse Family)

Identification: HB 10-13 cm, T 9-12 cm, W 30-80 g. The back of this mouse is covered in stiff bristles; the fur is grey-brown to sandy, white below; ears very large and long; tail long and almost hairless.

Distribution: Crete.

Habitat: Dry, rocky areas and scrub.

Behaviour: Active at dawn, dusk and by night; runs very quickly; lives in large families with a marked hierarchy; does not dig its own burrow but uses natural holes or rock crevices; females bring up their young communally.

Food: Seeds, fresh plant material, insects, molluscs.

Breeding: Mating season Mar-Oct; gestation 35-37 days; 2-5 litters per year, each with 2 or 3 relatively well-developed young which open their eyes at 2 days and leave the nest at 3 days old; young suckled for about 3 weeks.

The most striking feature of the Spiny Mouse are the stiff, almost hedgehog-like bristles covering the back. These bristles get longer towards the base of the tail and they are capable of being raised when the animal is alarmed. This is the sole European representative of a group of mice found mainly in southwest Asia and Africa.

Spiny Mice are well adapted to dry conditions; as long as their food has a relatively high water content they can even survive without drinking. Spiny Mice are born into a family group and mothers with young will readily suckle and groom each other's offspring.

Striped Field Mouse (above)
Spiny Mouse (below)

Northern Birch Mouse

Sicista betulina

(Jumping Mouse Family)

Identification: HB 5-7.5 cm, T 8-11 cm, W 5-13 g. Fur yellowish-brown with black traces and an obvious black stripe along the back; underside pale grey; tail slightly hairy and 1.5 times as long as the body.

Distribution: Eastern and northeast Europe, Scandinavia, Denmark; scattered in Germany and Austria. Not in the British Isles.

Habitat: Damp deciduous and coniferous woods with thick undergrowth, woodland clearings, raised bogs; mountains (particularly in the south of range – Germany and eastern Alps) to 2000 m.

Behaviour: Mainly active at dawn, dusk and by night; relatively tame; jumps and climbs well; builds spherical summer nest using plant material, usually anchored to a branch; hibernation (Sep-May) in a hole in the ground which it digs for itself.

Food: Seeds, berries, fruit, insects.

Breeding: Mating season May-June; gestation 28-35 days; a single litter each year with 2-6 naked, blind young; young open their eyes at 25 days and are suckled for 5 weeks.

The Northern Birch Mouse has an unusually long tail. Although they look rather like true mice such as the Wood Mouse and Harvest Mouse, the birch mice are actually related to Jerboas and Jumping Mice. Unlike all the other European rodent species, birch mice do not have a characteristic split upper lip. Like Harvest Mice, they use their long prehensile tail to help them as a support when clambering about in the undergrowth. Their fingers and claws are also very mobile and the outer small digit is opposable and can be used rather like a thumb to grip.

In the autumn, Birch Mice move from their spherical grass summer nest down into a burrow in the soil where they roll up into a tight ball and hibernate. They hold the record amongst mammals for length of hibernation, which can be up to eight months. They are also able to enter a state of lethargy during cold spells in the summer time, a behaviour otherwise known mainly in bats. Germany and Austria represent the most westerly and southerly outposts of their distribution.

A second birch mouse is also found in Europe, the **Southern Birch Mouse** (*Sicista subtilis*). This is found from Rumania eastwards, although it has also been discovered locally in the Hungarian lowlands and in southeast Austria (Burgenland). The main differences from the Northern Birch Mouse are the shorter tail and the clear white border on each side of the dark back stripe.

Northern Birch Mouse using its long tail as an anchoring device

Northern Birch Mouse: an attractive species rather rare in Europe

Crested Porcupine

Hystrix cristata

(Old-world Porcupine Family)

Identification: HB 60-80 cm, T 4-9 cm, W 10-15 kg. Bristly blackish-brown hairs with a white collar at the sides of the neck; hairs on back and tail modified as characteristic long black and white quills; quills at tip of tail further modified as a rattle; hairs on forehead and nape of neck long, forming a crest; ears rounded, almost covered by hair; front feet with powerful digging claws.

Distribution: Italy and Sicily; North Africa.

Habitat: Open woodland and scrub.

Behaviour: Nocturnal; lives in small family groups; either digs a burrow or occupies natural hole.

Food: Roots, bulbs, bark, fruit, occasionally insects and other small animals.

Breeding: Mating season May-Oct; gestation 110-120 days; litter size usually 1-3; young born with their eyes open and with quills (soft at first).

The Crested Porcupine was probably introduced early on (perhaps by the Romans) to Sicily and Italy from North Africa. It is a very large, impressive rodent, in Europe second in size only to the Beaver.

The quills, which can reach 40 cm long, are modified hairs, and are sharp, hard and pointed. During fights these loose quills can detach themselves and stick into the flesh of the opponent. When threatened, porcupines can rattle their quills and also impale them in the flesh of their enemies by reversing rapidly. The quills at the tip of the tail are shorter and hollow. These act as a rattle and thereby serve to warn off possible attack. During a concerted attack the Crested Porcupine will rattle its tail, spread its quills wide and run backwards towards its enemies making growling noises and also sometimes drumming with its hind feet. Crested Porcupines also occasionally use their spines in disputes with other porcupines. If a female is not willing to be mated she will use her spines to deter any persistent male. On the other hand, if she is receptive, she will signal this by raising her tail forwards over the back.

Newly-born porcupines also have spines. These are very soft to begin with, becoming hard in just a few days. Despite this, young porcupines are frequently groomed and licked by other members of the family group. Adult porcupines also indicate their interest in each other through licking. Porcupines have very hard, leathery tongues which protect them from their spines.

Tip of tail, showing rattle

Crested Porcupine with quills raised (above)
Porcupines often excavate burrows (below)

Brown Hare

Lepus europaeus

(Hare Family)

Identification: HB 40-70 cm, T 7-11 cm, W 2-7 kg. Fur yellowish-brown with grey tints, belly and underside of tail white; ears very long with black tips; hind legs much longer and more powerful than front legs.

Distribution: Whole of Europe except Iceland and northern Scandinavia; also absent from most of Iberia, Ireland and the Highlands of Scotland. Eastwards to Turkey and to central Asia.

Habitat: Originally open steppes, today mainly on cultivated treeless open fields; also found in swampy areas and in mountains up to 1500 m.

Behaviour: Most active at dawn, dusk and by night; runs with large bounds and on a very zigzag course when pursued; mainly solitary; den (also referred to as couch or form) is a shallow depression in the open or under grasses; involved in complex chases, especially during the mating season. This 'mad March hare' behaviour is usually unreceptive females chasing away males. Sometimes more than one male will follow a female and fights may then ensue.

Food: Vegetarian: grasses, herbs, shoots, berries, fungi, fruits, grain.

Breeding: Mating season throughout most of the year; gestation 42 days; 3-4 litters a year, each with 2-5 well-developed young which can run after just a few hours; young suckled for 3 weeks and independent at 4 weeks.

Hares and rabbits belong to an order called the lagomorphs and they are not closely related to rodents. Similarities include the possession of long incisor teeth for gnawing, but there are also many differences. In the upper jaw lagomorphs have two rows of incisors; they chew their food with a sideways jaw movement and never hold it between their front paws.

Destruction of forest and widespread development of open agriculture has increased the amount of suitable habitat for Brown Hares, which have now colonised much of Europe. Hares survive mainly through predator avoidance, for which they have several adaptations to help them. These include a keen sense of smell, large ears which can be moved individually, and large eyes, set wide apart on the head to give all round vision without having to move the head. By day Brown Hares often sit tight in their forms, with their nose facing the wind direction. They also sleep for very short periods of a few minutes or even just a few seconds at a time. (Continued p. 114.)

Track of Brown Hare (hopping)

Male Brown Hares in spring dispute over female (below)

Brown Hare (continued)

Adult Brown Hares occupy ranges of about 300 hectares, which they share with other individuals; they may move very long distances each day to their feeding grounds. They mark out regularly-used runways by leaving scent from their cheek and anal glands. They not only keep the grass down on these runways by the constant pressure of their feet, but also mow the grass by nibbling it short.

When a hare detects a potential enemy it does not run immediately, but lies flat against the ground with its ears down. In this position it relies on its good camouflage and often escapes detection. If however the predator gets too close the hare will suddenly jump up and race off. The top speed of a running hare is about 65 km per hour and it may take strides of more than 3 m at a time.

When two hares fight it is more likely to be over a female than over territory. Such fights between rival males may include bouts of boxing. They stand up on their hind legs and attack each other with their front paws, often pulling out tufts of each other's fur.

Female Brown Hares do not build nests in which to have their young but instead seek out a suitable dry, sheltered place in the open. Newborn Brown Hares have a full covering of fur and can see well as soon as they are born. In just a few hours they can walk and run. Young hares often crouch very still for long periods in the open and at this stage. Since they seem not to have developed their own characteristic body smell they are frequently overlooked by predators such as dogs or foxes. The mother hare visits each of her offspring about two or three times a day in order to feed them. After a week the young hares begin to nibble some plant food for themselves, and by three weeks are entirely weaned. Until they reach this age they are very sensitive to the cold and wet and many die during periods of bad weather. Unfortunately, many hares are also killed on the roads by traffic.

Hares boxing

Young Brown Hares lie out in the open (above)
Brown Hare in squatting position, showing camouflage (centre)
Brown Hare crossing a road (below)

Mountain Hare
Irish Hare
Tundra Hare
Lepus timidus

(Hare Family)

Identification: HB 50-65 cm, T 4-8 cm, W 2-6 kg. In summer, fur brown or greyish-brown with white belly and tail, in winter usually pure white; ears and tail shorter than in Brown Hare (p. 114); tips of ears always with black margins even when in winter coat; paws broad and hairy.

Distribution: Ireland, Scotland, northern England, Faroe Islands, Scandinavia, northeast Europe, taiga and tundra of northern Asia, North America, Greenland; also an isolated population in the Alps.

Habitat: Deciduous, mixed and coniferous woodland, moorland, hilly country up to high mountains.

Behaviour: Mainly active at dawn, dusk and by night; lives isolated or in small groups; makes form in old heather or between rocks and roots.

Food: Vegetarian: as for Brown Hare but heather is also important and bark in hard weather.

Breeding: Mating season Feb-Aug; gestation 42 days; 2-3 litters a year, each with 2-5 young; young born fully furred and with eyes open; development as for Brown Hare.

During the last Ice Age Mountain Hares inhabited the tundra-like ice-free zones. As the glaciers retreated, so they followed them northwards to Scandinavia, the north Atlantic islands and into the Alps.

The usual summer coat of the Mountain Hare is brown; this changes during the winter to pure white. The precise timing of this change depends upon the temperature. In the high north, Mountain Hares are white throughout the year, but in Ireland and the Faroe Islands for example, they remain brown even in the winter. In some parts of Scandinavia and Scotland the autumn colour change is incomplete and the animals have a mottled pattern through the winter.

The hairy feet of the Mountain Hare protect well against the cold of ice and frozen soil and act as snow shoes to help when running over snow cover. Mountain Hares are essentially animals of woodland, scrub and tundra. Where both hare species are found together they therefore do not usually compete. There are occasional reports of hybrids between the two species, but it seems that the offspring are infertile.

Mountain Hare: the ear tips are always dark brown or black, in summer and winter coat

Mountain Hare during coat colour changeover (above)
Mountain Hare in summer (below)

Rabbit

Oryctolagus cuniculus

(Hare Family)

Identification: HB 35-45 cm, T 4-8 cm, W 1-2.5 kg. Smaller than Brown Hare; fur sandy to dark greyish-brown, occasionally black; belly and underside of tail white; hind legs longer and more powerful than front legs; ears much shorter than those of Brown Hare.

Distribution: Western, central and southern Europe (except Balkan peninsula); northwest Africa. Widespread in Britain and Ireland, including most offshore islands.

Habitat: Warm, dry areas with sandy soils and low plant growth, heaths, pastures, parks, gardens; in hills not generally found above about 600 m.

Behaviour: Mainly active at dawn, dusk and by night; runs with a hopping motion or in long jumps, taking a zigzag course when fleeing; lives in colonies made up of several large families in burrow systems (warrens); warrens consist of many angled tunnels and a central living area; warns other rabbits in the colony of danger by thumping with hind feet.

Food: Vegetarian: grasses, herbs, bark, roots, bulbs.

Breeding: Mating season Feb-July, throughout the year in milder regions; gestation 28-31 days; 4-6 litters a year, with usually 4-7 naked, blind young; young open their eyes after about 10 days, are suckled for about 4 weeks and are then independent.

Note: This species is the ancestor of domestic rabbits.

Rabbits are warmth-loving creatures and at the time of the last glaciation were restricted to the Iberian Peninsula. Since Roman times their range has been extended and they have been introduced to many other regions either to provide food or for hunting.

Rabbits are sociable creatures, living in large family groups with a dominance hierarchy. The hierarchy is reinforced by ritualised fights in which the loser is not usually injured. On the other hand, fights between rival territorial groups of rabbits may result in bad wounds. Group territories are marked by rubbing landmarks (stones, logs) with scent from a gland under the chin.

Rabbit marking the border of its family territory with scent from the gland on its chin

Pregnant female rabbits dig out special nesting burrows which they line with grass and fur. Unlike young hares, baby rabbits are helpless and naked. When the female leaves the young she seals the entrance to the nest burrow with soil and grass, returning only once each night to suckle the brood for about five minutes. At four weeks old the young transfer to the family burrows and the female digs out a new nesting burrow for her next litter.

The very similar **Eastern Cotton-tail** (*Sylvilagus floridanus*) has been introduced to many parts of France, Spain and Italy for hunting. Unlike the Rabbit it does not dig extensive burrows.

Rabbits require a lot of warmth and often sunbathe (above)
Mother with young in nesting burrow (below)

Stoat

Mustela erminea

(Weasel Family)

Identification: HB 18-32 cm, T 9-14 cm, W 150-300 g; female smaller and lighter than male; body very slim and elongated; fur cinnamon-brown in summer with yellowish-white underside, usually pure white during the winter, but tail always yellowish towards the base and with a black tip.

Distribution: Arctic and temperate parts of Europe and Asia, also North America, widespread in Britain and Ireland. Absent from Iceland and the Mediterranean region.

Habitat: Wide variety, including woodland, scrub hedgerows and mountain scree, wherever they can find cover.

Behaviour: Active by night and day; characteristic bounding gait, often stands bolt upright; solitary except in breeding season.

Food: Small mammals (particularly rodents) and birds.

Breeding: Mating season May-July, but development of the embryos is delayed until the following spring when the young are born after an active gestation of 4 weeks (see p. 251 on delayed implantation); 4-8 blind almost naked young per litter, young open their eyes at 6 weeks and are independent at 3-4 months old.

The pure white winter coat of the Stoat (known as ermine) has long been used for making traditional ceremonial cloaks, often with a black tail tip sewn into a prominent position. One reason why ermine is so valuable is that it is unusually dense: 1 cm^2 of the pelt contains about 20,000 hairs. The black tail tip distinguishes the Stoat at all times of the year from the similar but smaller Weasel. However, not all stoats turn white in the winter – in the warmer parts of its range, including England, it may retain the summer coat throughout the year, or moult into a transitional patchy colouring. In any case there is a transition period which lasts for about a month, so in spring and autumn stoats often show intermediate colouration.

Stoats mainly hunt small rodents up to the size of a rat, although they will occasionally kill mammals as large as rabbits or even hares. On detecting their prey they creep as low as possible along the ground and then attack with a quick jump. The prey is then usually killed by a lethal bite to the back of the head or neck.

As with other members of the weasel family, the usual Stoat method of locomotion is by a series of jumps in which the back is strongly arched and the left and right feet make contact with the ground simultaneously. This gives a characteristic pattern of footprints.

The typical bounding gait leaves characteristic paired footprints

Stoat in summer coat (above left)
Stoat in transitional coat (above right)
Stoat in winter (below)

Weasel

Mustela nivalis

(Weasel Family)

Identification: HB 17-23 cm, T 3-6.5 cm, W 40-170 g; male somewhat larger and heavier than female; border between brown back and white belly fur along flanks usually rather irregular; tail brown and lacking black tip.

Distribution: The whole of Europe (except Ireland and Iceland); Asia and North America.

Habitat: Woodland, hedgerows, long grass, also found in cultivated areas and in mountains up to 3000 m. Tends to avoid damp ground.

Behaviour: Active by day and night, solitary. Moves in a typical series of short jumps and often stands bolt upright; climbs well. Hunts mice and often follows them into their burrows.

Food: Voles and other rodents, more rarely birds' eggs and insects.

Breeding: Mating usually occurs in spring and summer; 1-2 litters a year, each with 4-7 naked, blind young which are independent at 3-4 months old; male does not help rear the young.

Note: Smallest European carnivore.

The Weasel is a specialist hunter of small rodents such as voles and mice. With its narrow body and small head (the latter is only 2 cm wide) it can easily follow its prey along their burrows. The long thin body gives the weasel a relatively high surface-area to body-weight ratio, which means it loses a lot of heat. Its fat layer is also rather thin, so it must satisfy its high energy requirements with a large intake of food. For this reason weasels are nearly always on the hunt.

Each weasel has its own territory, the size of which is dependent on the density of small rodents but is usually between five and eight hectares. In a single night a Weasel can travel up to the equivalent of about 2 km. Although Weasels occasionally eat birds, frogs and even young rabbits, their specialisation on mice and voles makes them extremely useful in gardens and other cultivated areas.

The nest is carefully lined with hay and moss and is often constructed in an old vole or mouse burrow. Young Weasels grow very rapidly and can even produce a litter of their own in their first year. This short generation time means that Weasel populations can increase very rapidly in response to a population explosion in the vole community. During times of such plentiful food a female Weasel can produce litters of up to 12 young. Weasels can also breed at any time of the year, and pregnant females have been found in every month.

In the north of its range, and also in high mountains, Weasels turn white in the winter but they remain brown in the milder parts of their range, including the whole of Britain. Weasels however never have a black tip to their tail and this means that they can be easily distinguished from stoats at all times of the year.

The Weasel, unlike the Stoat, has a wavy border between the brown back and white belly

European Mink
·Mustela lutreola

(Weasel Family)

Identification: HB 32-40 cm, T 13-18 cm, W 550-800 g; ears short and rounded; tail less than half of the body length; fur dark brown; lips and chin white.

Distribution: Eastern Europe and Asia, with isolated populations in western France and northern Spain.

Habitat: Near slow-moving fresh water.

Behaviour: Active by night and day, solitary; swims and dives well; lives in a river bank in tunnels which it may dig for itself. Lays up food stores during times of plenty. Does not hibernate.

Food: Rodents, frogs, fish, water birds, eggs.

Breeding: Mating season April, gestation 6 weeks, 3-7 young which open their eyes at 30 days old, are suckled for 2 months and are independent at 3 months.

Note: Vulnerable as a wild species.

In the last century the European Mink was found over most of Europe. However, it has been exterminated over most of its range, except in the far east of Europe and one or two pockets in western France and Spain. In habits and feeding biology, mink are intermediate between martens (p. 130, p. 132) and Otter (p. 138).

Mink hunt mainly by swimming and diving, but they are not as adept aquatic predators as Otters and seldom catch large, healthy fish. Typical mink hunting territory is a thickly wooded river bank where the current is fast enough to retain some open water even in winter, and it usually covers about 15-30 hectares. The tunnel system normally has an entrance near the water and also an escape route on the landward side for emergency use during times of flood.

For a long time mink were thought to be nocturnal animals. However, they are also active during the day and they have good colour discrimination in their vision. They are however extremely shy and this, combined with their preference for rather impenetrable habitats, means that they are rarely seen by people during the day, even in areas where they may be common.

American Mink (*Mustela vison*) is very similar in appearance. This species was introduced to several parts of Europe, including the British Isles, Iceland and eastern Europe, often as an accidental escape from fur farms. In some areas it has increased so rapidly that it threatens the native populations of European Mink.

European Mink (right) showing white upper and lower lips. American Mink (left) has white only on the lower lip.

Mink showing silky gloss to the fur (above)
American Mink (below)

Western Polecat

Mustela putorius

(Weasel Family)

Identification: HB 35-46 cm, T 12-17 cm, W 600-1500 g; female smaller and lighter than male; fur brownish-black with yellow ochre underfur on the back; face pale with dark eye mask; ears short and rounded with white edges; tail about a third the length of the body.

Distribution: The whole of Europe except Ireland, Iceland, northern Scandinavia and the Mediterranean islands; east to Ural mountains. In Britain exterminated by the early 20th century except in Wales, but now recolonising adjacent parts of England.

Habitat: Mainly mixed landscape with meadows, fields and scattered woodland, river valleys, often found close to inhabited areas.

Behaviour: Mainly active at dawn, dusk and by night; solitary. Produces foul-smelling secretions from the anal glands, used in marking territory and also defence. Moves mainly by jumping.

Food: Small rodents, frogs, birds, sometimes snakes and insects.

Breeding: Mating season Mar-May; gestation 6 weeks, 3-7 young per litter; these have pale coats and open their eyes at a month old; by the time they are independent in the following autumn their coat colour has turned brownish.

The Western Polecat is predominantly nocturnal. By day it usually sleeps in a tunnel or beneath brushwood and leaves in a tree stump or hollow branch. At night polecats set out on their nocturnal hunting forays. The hunting territory can be several square kilometres in extent. Each polecat marks the borders of its territory using a pungent secretion from the anal glands. A cornered polecat often squirts this foul-smelling substance at its enemy in self defence.

A male polecat will often follow a female for long periods in the mating season. If the female is receptive, the male grabs her by the neck and drags her around, a behaviour which apparently stimulates ovulation.

The domesticated form of the polecat is the **Ferret** (*Mustela furo*). This is sometimes pure white with red eyes, but often cream-coloured or light brown showing a 'dilute' polecat pattern. Ferrets are sometimes trained and used for hunting rabbits. The ferret will enter a rabbit warren and chase the rabbits out of their burrows. If a ferret catches a rabbit below ground it may eat it there and then and fall asleep. For this reason working ferrets are often fitted with small muzzles to prevent them attacking the rabbits underground.

Ferrets also make quite good, if somewhat smelly, pets since they are very tame. In some areas they have established wild populations and may also produce ferret–polecat hybrids. Hybrid animals tend to be paler than true polecats and lack the characteristic face mask.

Young Western Polecats (above)
The Western Polecat is rarely seen during the day (below)

Steppe Polecat

Mustela eversmanni

(Weasel family)

Identification: HB 30-56 cm, T 8-18 cm, W 1-2 kg; identical in form to the Western Polecat (p. 126) but fur pale yellowish-brown except for contrasting dark mask, legs and underside.

Distribution: Eastern Europe from Austria to southern Russia and east as far as China.

Habitat: Open grassland.

Behaviour: Nocturnal. Lives in burrows, often old souslik or hamster burrows.

Food: Mainly rodents, including sousliks, hamsters, voles and mice.

Breeding: Mating season Feb-Mar; gestation 5-6 weeks, litter of usually 3-6 (although sometimes many more) born in Apr-May. Weaned at 6 weeks and become fully independent at about 3 months old.

Where the ranges of the Western and Steppe Polecat overlap they do not compete since they occupy rather different habitats: the Western Polecat prefers wooded country and is also found close to human habitation, whereas the Steppe Polecat is an animal of open, treeless steppe, far from towns and villages. Steppe Polecats produce particularly large litters – up to 18 blind young can be born, usually in an old souslik (p. 62) burrow. The young have pale fur at first but by the time they emerge from their burrow at about a month old their eyes are open and their fur has turned dark. By late summer they leave the family burrow and are independent.

The Steppe Polecat (right) is noticeably paler than the Western Polecat (left) and it also lacks the dark facial mask

Marbled Polecat

Vormela peregusna

(Weasel Family)

Identification: HB 27-35 cm, T 12-20 cm, W 370-715 g; body fur light yellowish-brown with dark mottling; legs and belly black, dark mask on face; ears large and broad with a white margin; tail bushy.

Distribution: Southeast Europe, western Asia as far as western China.

Habitat: Treeless steppe and semi-desert.

Behaviour: Mainly active at dawn and dusk, solitary; does not dig its own burrows but enlarges tunnels of steppe rodents, e.g. sousliks or hamsters.

Food: Rodents (e.g. hamsters, sousliks), more rarely birds and reptiles.

Breeding: Mating season Feb-Mar; gestation 8 weeks; 3-8 young; development as for Western Polecat.

This attractively marked species is becoming steadily rarer as its steppe habitat disappears. Like other members of the weasel family it moves in a series of forward jumps interspersed by standing bolt upright. Hunting is mainly below ground where it chases hamsters (p. 66) and sousliks (p. 62) in their burrows. When threatened, the Marbled Polecat fluffs up its tail and arches it over its back in a squirrel-like posture. It also kicks out with its hind legs, grinds its teeth and howls like a dog. If the threat continues it can also release a foul-smelling secretion.

Steppe Polecat (above)
The Marbled Polecat has highly characteristic markings (below)

Pine Marten

Martes martes

(Weasel Family)

Identification: HB 40-55 cm, T 20-28 cm, W 900-2000 g; colour uniform chestnut-coloured to dark brown; chin and breast pale yellow or orange-yellow with darker border; ears large and rounded, yellowish inside and around rim; tail long and bushy.

Distribution: Europe, except Iceland and southern Spain; Caucasus, western Siberia. Present in Ireland and upland parts of Scotland, northern England and Wales.

Habitat: Wooded country, in mountains up to the tree-line.

Behaviour: Mainly active at dawn, dusk and by night, solitary; climbs exceptionally well but also hunts on the ground; lives in holes in trees, sometimes in squirrel dreys or old birds' nests.

Food: Mice, squirrels, birds, insects, fruit.

Breeding: Mating season July-Aug; gestation extended to about 9 months by delayed implantation (see p. 251); the 3-5 blind young are born in April and open their eyes at 5 weeks; they can climb and jump by about 8-10 weeks old and by the summer they are independent.

Unlike the closely related Beech Marten (see p. 132) Pine Martens are not usually found close to human habitation, but they can be attracted to bird-tables in rural gardens. They are shy of people and have become rather rare over much of their range. The main threats have been the destruction of virgin forest habitat and they have also long been hunted for their valuable pelts.

Pine Martens are exceptionally good at climbing – they can jump from branch to branch and run down as well as up tree trunks. Their speed and agility in the trees means that they can catch squirrels as well as voles and small birds.

In the mating season (high summer) Pine Martens make shrill cat-like calls. After mating, the male and female go their separate ways again. Implantation of the fertilised eggs is delayed until the start of the following year.

Pine Marten (left) and Beech Marten (right) are mainly distinguished by the shape and colour of the throat markings

At that time the embryos begin to develop and the young are born in a hollow tree in April. Young animals already show the characteristic pale neck markings.

The related **Sable** (*Martes zibellina*) is now extinct in Europe and is found only in Siberian forests. It resembles the Pine Marten but has somewhat longer legs and larger ears. Sable produces the most sought after fur. The species has been reintroduced to many parts of Russia and Siberia for this reason.

A marten relaxes during its midday rest (above)
Anxious watchfulness (centre)
Threat (below)

Beech Marten
Stone Marten

Martes foina

(Weasel Family)

Identification: HB 43-55 cm, T 22-30 cm, W 1200-2000 g; similar in shape to Pine Marten but has shorter legs; throat and breast white, with bib often divided into two by a dark central stripe; ears with white border; chattering voice.

Distribution: Southern and central Europe, Turkey and central Asia. Absent from British Isles and Sardinia.

Habitat: Rocky country, woodland margins, open montane forests; also found in villages and cities, parks and quarries.

Behaviour: Mainly nocturnal, solitary; does not climb as well as Pine Marten; stores food in the ground; lives in rock clefts, caves, buildings, ruins etc.

Food: All sorts of small animals up to the size of rabbits and hens; often takes mice and rats; also eats fruit and berries.

Breeding: Mating season July-Aug; gestation extended through delayed implantation to 8-9 months (see p. 251); 3-7 young, for development see Pine Marten (p. 130).

Beech Martens are not as agile in trees as Pine Martens. Yet on the ground they are just as fast even though they have shorter legs and are somewhat heavier.

They are very useful in villages and towns because they catch a lot of mice and rats. However they do occasionally cause damage, for example when they get into hen coops or pigeon lofts. In wilder areas Beech Martens will make stores of dead prey and they will often return to sites which have provided a good source of food. When the supply of fresh food is exhausted they fall back on their stores. Multiple killings, for example in a hen pen, may happen when a Beech Marten has entered it through a very narrow crevice and is unable to get out again carrying its prey. Under these conditions it will instinctively drop the first prey item and kill again. When no more hens are left alive it then begins to eat and may even fall asleep at the scene of the carnage. In the wild such scenes are rare as the potential prey species can usually escape.

The mating season is in high summer but those females that are not impregnated at this time of year pair again in the following February or March. In either case the young are born in late spring. In summer matings, the implantation of the fertilised eggs is delayed and in this case gestation lasts for eight to nine months.

In some parts of central Europe Beech Martens have been known to cause damage by biting through cables in parked cars. It seems that this may be the work of young, inexperienced animals which tend to bite any unusual item to test for edibility.

Although they can climb well Beech Martens spend most of their time on the ground (above)
Note characteristic divided white bib marking (below)

Wolverine
Glutton

Gulo gulo

(Weasel Family)

Identification: HB 70-90 cm, T 18-23 cm, W 10-20 kg; male larger and heavier than female; largest member of the Weasel family, resembling a small brown bear but with a bushy tail; long-legged, with broad paws and powerful claws.

Distribution: Taiga and tundra from Norway to Russia, northern Asia and North America.

Habitat: Open coniferous woodland and forest tundra, in mountains up to the tree-line, moorland; avoids cultivated areas.

Behaviour: Active by day and night, solitary; climbs well but lives mainly on the ground; territories can be up to 2000 km^2; does not use a permanent den during winter; breeds in caves in the ground, in rocks or beneath the snow.

Food: In summer takes eggs, birds, lemmings, berries; in winter Reindeer, Roe Deer, Mountain Hares, also carrion.

Breeding: Mating season Apr-June; gestation (including delayed implantation, see p. 251) about 9 months; 2-3 blind young which open their eyes at 5 weeks, are suckled for 3 months and remain with their mother for at least a year.

This rather bear-like mammal is a close relative of the Martens. Like them it also has paired anal glands which produce a foul-smelling secretion. A further scent gland in the genital region is used to mark out the territory. The territory of a male commonly includes those of several females. In areas where prey animals are scarce a Wolverine's territory can encompass as large an area as 2000 km^2 and the animal may cover up to 40 km a day. Its broad foot pads enable it to move quickly even over loose snow and thus gain advantage over its prey.

Wolverines can kill prey as large as Reindeer or Elks, although with such large prey animals it is usually the weaker or ill individuals that are taken. They also eat a lot of carrion and sometimes rob other animals of their prey (and may even steal from hunters' cabins). Large prey items are often divided up and the remnants stored for use in times of food scarcity.

Wolverines have been hunted for their fur which is used particularly to trim parkas. However in fur-trapping areas it has also been persecuted because of its habit of emptying traps before the trappers reach them.

Despite its weight and somewhat dumpy body, the Wolverine can climb well

Wolverine on the look out (above)
Stretching and yawning soon after waking up (below)

Badger

Meles meles

(Weasel Family)

Identification: HB 60-90 cm, T 12-24 cm, W 7-13 kg in summer, 16-24 kg in autumn; black and white stripes on the head, ears with white border, legs and belly black, back silvery-grey; front feet have long claws.

Distribution: Whole of Europe except northern Scandinavia, Iceland, Corsica, Sardinia, Sicily and Cyprus; also in temperate parts of Asia, east to China.

Habitat: Deciduous and mixed woodland, hedgerows, gardens, parks.

Behaviour: Most active at dawn, dusk and by night; uses regular pathways and latrines; digs extensive burrows, often occupied by a small group of 5-10 individuals with one dominant boar. Activity reduced during hard winter weather but no true hibernation.

Food: Omnivorous; food includes earth worms (often the dominant item), many kinds of insects, frogs, small mammals, carrion, roots, buds, fruits and berries.

Breeding: Mating season Feb-Oct, mostly Mar-May; gestation variable according to length of the delayed implantation period (see p. 251), the development of embryo takes 60 days; 1-5 (usually 2) blind young with pale fur; these young open their eyes at 30 days and are suckled for about 3 months.

Badgers are mainly nocturnal. They usually forage individually, following well-used tracks and digging in the soil from time to time. They are omnivorous and somewhat opportunistic about what food they will take. This can include earthworms, beetles, voles and other small mammals, carrion, roots, fruit and other plant material.

Their powerful front feet with long claws are ideal for digging in the soil and they use them to create extensive underground burrows called setts. A central nesting chamber is usually excavated 5-10 m from the entrance and about 3 m deep. This is lined with dry vegetation and a new one is dug for each litter. Large burrow systems may be occupied for many generations and usually have several entrances and emergency exits. There may be up to 40 entrance holes and the burrow systems may extend up to 100 m.

In cold snowy winters Badgers enter a period of dormancy, although they do not truly hibernate. Under these conditions they may remain in their burrows for weeks, living off their fat reserves.

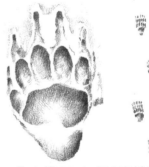

Track of Badger, showing marks left by long claws. Each mark is effectively a double footprint since they usually place their hind feet directly on the print left by the front feet

Badger near burrow entrance (above)
Half grown young Badgers (below)

Otter

Lutra lutra

(Weasel Family)

Identification: HB 55-90 cm, T 30-50 cm, W 5-10 kg; male larger and heavier than female; fur shiny brown, yellowish-brown round chin and neck (sometimes whitish); head flat; ears small and rounded; stiff whiskers; tail long, thickening towards base; feet webbed and clawed.

Distribution: Europe (except Iceland and the larger Mediterranean islands); Asia; northwest Africa. This species is vulnerable and has become extinct or reduced in numbers over much of its range.

Habitat: All kinds of freshwater, usually with thick bank vegetation; also on some secluded rocky coasts.

Behaviour: Mainly nocturnal but active by day where undisturbed; solitary; swims and dives exceptionally well; territorial; nests in cavity in soil or a hollow tree in riverside vegetation; entrance usually below water.

Food: Fish, also locally Muskrat, water birds, frogs, crustaceans.

Breeding: Mating season Feb-Apr, sometimes throughout the year; gestation 62 days; 2-3 blind young with greyish hair; these open their eyes at 5 weeks, can swim at 10 weeks and first leave their family in the following spring.

Note: One of the most threatened European mammal species.

Otters are somewhat ungainly when moving on land but are superbly adapted to life in the water, being excellent swimmers and divers. For slow swimming they use their webbed feet, the toes of which can be spread very wide. When swimming fast they move by swaying their whole body and tail from side to side, trailing their front feet next to the body. Both ears and nostrils close when diving. Otter fur has an oily secretion covering it which makes it waterproof. The thick underfur retains a layer of air which simultaneously prevents water-logging and also keeps the body warm.

Otter territories vary quite a lot in size, mainly dependent on the supply of fish. Otters have to eat about 12-15% of their body weight in food per day, that is about 1-1.5 kg; their territories may stretch for up to 30 km on each side of a river. Otters normally hunt during the night, diving after fish and engaging in short underwater chases. Small fish may be eaten in the water and the Otter may turn over onto its back and manipulate its prey with its front feet. Larger prey are always taken to the bank before being eaten. Otters nearly always leave the water at the same place and they leave characteristic piles of droppings (spraints) at particular points along the track. On land they follow a network of frequently used trails.

(Continued p. 140).

Otter standing alert using its tail as a strut

Otters at water's edge (above)
Carrying fish to bank (below)

Otter (continued)

Otters lead a mainly solitary life and have a short mating season. This usually takes place in March (although they do sometimes mate at other times of the year) and is accompanied by repeated whistles and wild water games.

The mouse-sized young Otters are grey in colour and are born in a burrow lined with soft material. They stay with their mother for a long period and gradually learn to swim, dive and hunt. Young Otters indulge in a lot of play activity in preparation for adult life. They sometimes use slippery clay banks or icy slopes as slides into the water. When they are about a year old and independent they often have to travel a long distance to find a territory of their own and at this stage may be found a rather far away from the water.

Otters are easily tamed and in parts of Europe they have been used to help fishermen, a custom that still persists in some regions of China. Conversely, Otters are often regarded as competitors by fishermen and have been much persecuted for this reason, as well as for their pelts. This persecution, combined with the effects of pollution, habitat destruction and disturbance, have led to extinction over much of its former range. Pollution of water by detergents and other chemicals can destroy the waterproofing of the Otter's coat, often with lethal result. Although Otters are now protected in almost all European countries it is doubtful that they will be able to recolonise all their former habitats.

Otters declined drastically in lowland Britain during the 1960s, probably due mainly to the accumulation of pesticides in their prey. Since then there has been a slow recovery, helped by some reintroductions, but they are now numerous only on the coasts of western Scotland and the Scottish islands.

Otters have webbed paddle-like feet.
In soft sand or mud Otter prints leave a
highly characteristic wavy line, left by the tail

The otter is water-proofed by an oily secretion on the coat

Raccoon

Procyon lotor

(Raccoon Family)

Identification: HB 50-70 cm, T 20-30 cm, W 5-10 kg; fox-sized, rather squat animal with grey-brown fur and a black and white face mask; tail bushy, with 5-7 dark rings.

Distribution: North and Central America. Deliberately introduced or escaped from fur farms in Germany and Russia, now spread as far as northwest France and Holland.

Habitat: Deciduous and mixed woodlands, parks, particularly near water.

Behaviour: Nocturnal, solitary; climbs and swims well; breeds in hollow trees and also spends the days there, often 10-12 m high; uses front feet to gather food; has a period of winter dormancy in colder regions.

Food: Mainly aquatic animals, insects, worms, birds' eggs and nestlings; in autumn mainly berries and fruits.

Breeding: Mating season Jan-Feb; gestation 63 days; 2-5 blind young with thin fur which open their eyes at 22 days and are independent at 6 months.

Note: Albino and melanic forms are occasionally seen.

Raccoons use their front paws like hands to find and manipulate food. They turn over small stones in shallow water in their search for food, often watching elsewhere as they do so. This behaviour is instinctive, and even captive animals tend to put their food into water and then remove it again. This habit has led to the erroneous belief that raccoons habitually wash their food before eating it.

Front paw of Raccoon showing long, flexible fingers (left); trail left by running Raccoon (right)

In spring and summer they feed mainly on earthworms, slugs, snails and other invertebrate soil and water animals. They seldom take larger prey and therefore pose no threat to wild game. They will occasionally plunder birds' nests, but the extent of this has been exaggerated. Raccoons are attractive animals, originally introduced to the continent for their fur, and are now well established over much of northern central Europe.

In the autumn, Raccoons switch their diet to include acorns and other fruit, and at this time of year they lay down a thick layer of fat to last them through the winter months. In extended cold periods they enter a state of dormancy, usually in tree-holes.

Raccoon showing characteristic searching behaviour in shallow water

Brown Bear

Ursus arctos

(Bear Family)

Identification: HB 1.7-2.2 m, T 8-10 cm, W 100-300 kg; fur pale brown to almost black, young animals often with a yellowish collar; eyes small; ears rounded and with a thick covering of hair; tail hidden by body fur; powerful unretractable claws.

Distribution: Scandinavia, Balkans, Russia; also relict populations in the Pyrenees, northwest Spain and central Italy; Asia, North America. There are some 30 subspecies of the Brown Bear, including the larger Grizzly and Kodiak Bears in North America.

Habitat: Undisturbed, well-wooded country.

Behaviour: Active by night and day; normally solitary, but will gather together in groups to exploit rich sources of food; individual territories overlap; swims and climbs well; enters a period of dormancy during the winter; young born in winter quarters.

Food: Omnivorous: berries, roots, shoots, honey, ants, insect larvae, fish, birds, eggs, carrion, occasionally large mammals such as deer.

Breeding: Mating season Apr-July; gestation 6-7 months including delayed implantation; 2-3 blind young which open their eyes at 4-5 weeks, are suckled for 4 months and remain with their mother until they are about 2 years old.

Historically, bears and people have long been in conflict and bears have disappeared from all the well-populated areas of Europe. When people first began to make inroads into the natural wild forests of Europe, bears gradually got a reputation for raiding villages in search of fruit or domesticated animals and this set up a traditional enmity. This interaction also gave rise to many stories about the extraordinary strength of the Brown Bear, leading to a bear cult. Today bears feature widely in heraldic crests as symbols. Bears are intelligent animals and can be tamed and taught tricks; they are often used in circuses, and travelling folk used to display dancing bears.

The largest Brown Bear subspecies are the North American Kodiak Bear and the Kamchatka subspecies from northeast Asia. European Brown Bears by contrast are much smaller. In southern Europe for example they are only about 1.7 m long and weigh about 200 kg. The largest specimens in Europe are found in Scandinavia and Russia. These reach about

Comparative sizes of the European Brown Bear (foreground) and the Kodiak Bear, a North American subspecies

2.2 m and can weigh nearly three times as much. This variation follows a general biological rule whereby animals of the same or closely-related species tend on average to be smaller in warmer areas. (Continued p. 146).

Young Brown Bears must practice to develop their climbing skills

Brown Bear (continued)

In the breeding season male Brown Bears follow the females over long distances. The males rarely fight over the females and it is not necessarily the strongest male that succeeds. In December or January, two or three small (about rat-sized) young are born in the winter den. After mating, the development of the embryo is held up at a fairly early stage and gestation proper begins in November, taking another eight to ten weeks. Similar cases of delayed implantation or delayed embryonic development can be found, for example, in Roe Deer, Stoats and Badgers.

The mother Brown Bear takes a lot of care of the young cubs and protects them from danger; they remain with her for two years. During this period female bears can even be a threat to people, so strongly do they protect their young, although Brown Bears are usually very shy and go out of their way to avoid any contact with humans. Near inhabited areas Brown Bears are almost entirely nocturnal.

Mother bears sometimes suckle their young in a sitting posture

A bear's jaws mirror its omnivorous diet by combining typical carnivore features (powerful canines) with specialist vegetarian teeth (flat-crowned molars). The latter help bears to grind up their plant food. Although often largely vegetarian, Brown Bears are quite capable of overpowering sheep, goats and deer. Many bears specialise on plants or insects however. Such populations are known in Sweden as Grass Bears and in Russia as Ant Bears.

In harsh winter weather bears retreat to rocky dens or holes in the ground which they line with moss and grass. Here they live mainly off their fat reserves which they lay down in the autumn. This is not however a true state of hibernation and their body temperature, heart rate and breathing rate are comparable with those during normal sleep. Dormant bears are therefore ready at any time to respond to danger and change their winter quarters should the need arise.

Skull of Brown Bear showing typical omnivore dentition

Brown Bear adult in characteristic posture (above)
Young Brown Bears at play (below)

Polar Bear

Ursus maritimus

(Bear Family)

Identification: HB 2-2.5 m, T 8-10 cm, W 150-600 kg, in winter males up to 800 kg; female smaller and lighter than male; fur pure white to straw coloured; eyes small; lips and tip of nose black; tail covered by body fur.

Distribution: Arctic coasts and seas. Resident around Spitzbergen, and along the coasts of Siberia, North America and Greenland. Polar Bears occasionally reach the north of mainland Norway and Iceland.

Habitat: Ice packs, islands and coasts around Arctic Sea.

Behaviour: Diurnal; solitary; wanders long distances or moves around on pack ice; swims and dives exceptionally well; pregnant females overwinter in dens they dig for themselves in mounds of snow.

Food: Seals, fish, carrion.

Breeding: Mating season Mar-Apr; gestation 8-9 months, extended by delayed implantation; 1-3 (usually 2) tiny blind young which open their eyes at a month old; young suckled for 18 months and independent at 2 years.

Note: Largest living land carnivore.

Polar Bears are exceptionally well-adapted to the open ice fields of their natural habitat. They have very thick fur that insulates them well from the cold of their environment and is particularly water-resistant. They also have a thick fat layer beneath the skin and even their hairy feet are well protected by a thick layer of fat.

Polar Bears are always on the move and wander over very large distances. They are also often moved passively by the natural drifting of pack ice, often for hundreds or even thousands of kilometres.

Only the females show sedentary behaviour, when they are bringing up their young. Pregnant females dig out a nesting chamber at the beginning of winter, usually in a sheltered site in deep snow. The warmth of her body soon creates an inner icy layer on the walls of the den and the entrance may be sealed with a layer of snow. Here the rat-sized completely helpless young Polar Bears are born in

Cross-section of maternity den in a snow bank. The entrance is blocked by a layer of snow during the winter

December or January. For the first few weeks the mother keeps the new born cubs warm by lying on her back or side in the nest chamber and cradling them on her belly. They develop very quickly and are ready to follow their mother out of the den in March or April. The cubs stay with their mother for about two years so adult female Polar Bears usually produce a litter every third year. This relatively slow reproductive rate makes the species particularly vulnerable to over-hunting.

The major prey species of the Polar Bear are seals, particularly Ringed Seals and Bearded Seals. Seals may be ambushed on their way back to the water or caught at breathing holes in the ice.

Female Polar Bear with half-grown young (above)
Adult Polar Bear on pack ice (below)

Common Genet
European Genet

Genetta genetta

(Civet Family)

Identification: HB 40-55 cm, T 38-48 cm, W 1-2.3 kg; about the size of a domestic cat but with a more elongated, slender body and relatively short legs; fur yellowish or brownish-grey with lines of dark spots; tail long and bushy with 8-10 black rings; white spots beneath the eyes and at the sides of the nose.

Distribution: Iberian Peninsula, southern France, Balearic Islands, Africa, southwest Asia.

Habitat: Dry savannah, open scrub and rocky areas up to 2500 m.

Behaviour: Nocturnal; solitary outside breeding season; climbs well, downwards as well as upwards; creeps silently when hunting; repeatedly grooms itself by licking.

Food: Small animals, particularly mice and other rodents.

Breeding: Mating season Feb-Mar and July-Aug; gestation 10-11 weeks; 2-4 young are born furry but are blind for the first 8 days; they are suckled for 6 months and stay with their mother until they are about a year old.

The Common Genet is related to mongooses and is a representative of one of the oldest mammalian families. Like most nocturnal mammals it has unusually large eyes which enable it to pick up movements of its prey even in very low light. Also, like many nocturnal animals it is colour-blind.

This attractive and elegant animal has brought the art of creeping to a state of perfection; genets can glide soundlessly even through thick undergrowth. They are also adept at climbing trees and, like squirrels, can climb downwards as well as up. With a sudden forward spring they catch mice and other rodents using their front paws. Other cat-like features genets display are their retractable claws and the almost obsessive grooming of their fur.

Genets were often tamed in southern Europe for their mouse and rat-catching ability, although after the Middle Ages they were gradually replaced by the familiar modern house cat.

Face of Common Genet showing characteristic black stripes and white triangular marks under the eyes.

Most members of this family produce a musky secretion from their scent glands. This substance (known as civet) was treasured as an essential raw material for the perfume industry and was very expensive. Nowadays however it is of no particular commercial value.

Common Genet resting in a tree (above) and on the ground (below)

Egyptian Mongoose

Herpestes ichneumon

(Mongoose Family)

Identification: HB 57-65 cm, T 45-50 cm, W 1.9-4 kg; fur grey-brown, speckled, rather rough; head pointed; legs short and black; tail long with thick hair at the base and tapering to a sharp point.

Distribution: Southern Spain and Portugal; Dalmatia (introduced); Africa.

Habitat: Thick scrub, rocky country.

Behaviour: Diurnal, usually solitary, occasionally in family groups; lives on the ground; creeps silently and when hunting can make rapid changes in direction and lightning fast jumps; rests in holes in the ground or rocky crevices.

Food: Small animals up to the size of rats, often takes snakes and lizards, eggs and fruit.

Breeding: Mating season Mar-Apr and July-Aug; gestation 9-10 weeks, 2-4 young; for development see Common Genet (p. 150).

This species was given religious status by the ancient Egyptians and is depicted in many of their wall paintings and frescoes. It was often kept in a semi-domesticated state in villages for its mouse and rat-catching abilities. When hunting, the Egyptian Mongoose creeps quietly with its body pressed close to the ground and its feet almost invisible, making use of all available cover. At the same time it repeatedly tests the air with its delicate nose. It then captures its prey with a sudden rapid jump. If the prey is a snake the mongoose can avoid being bitten with its agile body movements and by making unbelievably rapid changes in direction. It can jump straight upwards into the air, roll over and over or even run backwards. When attacked, mongooses can raise their body fur thus almost doubling their apparent size. This behaviour also means that snakes often find themselves biting into the fur rather than the mongoose's body. Mongooses are not immune to snake poison although they can tolerate a significantly higher dose than many other mammals.

The Egyptian Mongoose has needle-sharp teeth.

Mongooses often travel in friendly groups in the summer, male and female leading the youngsters behind them. The young stay with their mother for about a year after birth.

The somewhat smaller **Indian Grey Mongoose** (*Herpestes edwardsi*) was introduced to Italy in 1955, to the area just south of Rome. This species is famous for its snake-killing prowess, even highly poisonous and dangerous species such as cobras. Another species, the **Small Asian Mongoose** (*Herpestes javanicus*) has been introduced to some of the Adriatic islands of Croatia.

Mongooses are not particularly adept at tree-climbing being much more at home on the ground (above)
Note the silvery-grey guard hairs covering the grey-brown coat (below)

Wolf

Canis lupus

(Dog Family)

Identification: HB 1-1.6 m, T 30-50 cm, W 30-75 kg; resembles a large rather heavily built German Shepherd Dog (Alsatian); ears large and triangular; eyes appear somewhat narrowed and slanting; tail bushy; colouring very variable, from grey through grey-brown to black; chin and throat usually paler; tip of tail black; produces characteristic long drawn-out howl.

Distribution: Once found across the whole of Europe but today only remnants of the former population exist, mainly in eastern Europe, scattered in Scandinavia, and mountains of southeast Europe, pockets in Spain, Portugal, Sardinia and Italy; Asia; North America.

Habitat: Originally in a wide range of habitats, today tends to be found mainly in inaccessible mountain areas.

Behaviour: Active by day and night although mainly nocturnal in Europe; hunts in packs outside the breeding season; highly developed social life; often pair for life.

Food: Medium-sized to large mammals, predominantly deer, sometimes smaller animals such as rabbits, hares and birds; also carrion and will sometimes scavenge at rubbish dumps.

Breeding: Mating season Dec-Mar; gestation 9-10 weeks; usually 4-6 blind, grey-brown young per litter. Pups open their eyes at 10 days old, are suckled for 2 months and are independent at about 6 months; male helps to rear the pups.

The wolf is the ancestor of all breeds of domestic dog and it exhibits many of the behaviour patterns familiar to us from dogs. However, in the wolf, barking is restricted to a short 'woof', usually employed as a warning signal. The characteristic howl is usually used for long-distance communication. Wolves sometimes howl in chorus and can be heard over long distances by other members of the pack or those of neighbouring territories.

The wolf pack is a highly developed social system and within the pack there is a hierarchy dominated by one breeding pair. Within this social structure serious fights between pack members are rare. Lower ranking pack members display a submissive gesture to those higher up the hierarchy. This involves presenting their unprotected neck, which instinctively causes the higher ranking wolf to refrain from attacking (see picture overleaf). The strongest male wolf is the leader of the pack, although on some hunting forays the highest ranking female may take charge. (Continued p. 156).

Wolves communicate by facial expressions.
Snarl preparatory to attack (left);
submissive expression (right).

Female with cubs (above)
Pack members of the look out (below)

Wolf (continued)

Wolves nearly always hunt in packs which vary in size (from fewer than 10 to about 30) depending on the size of the main prey animals. The pack will either follow the prey in a broad front or separate and close in from either side. Unlike cats which use a creep and jump method when hunting, wolves hunt on the run and can keep up a good speed for a long time. After a hunt of over an hour even the most powerful deer will tend to tire. At this stage the wolves close in simultaneously for the kill. A wolf pack is quite capable of overpowering even a fully grown Elk which would be quite impossible for a single wolf. The food is divided up

Submissive (left) and dominant postures shown by two members of the pack

amongst all the pack members. In areas where wolves have access to them they do take a certain number of domesticated animals such as goats and sheep, as these are easier for them to catch than truly wild ones. For this reason wolves have long been hunted and exterminated from much of Europe. They pose no threat to people however.

The mating season begins in the middle of winter. Wolf pairs are faithful to each other for long periods, often even for life, and at mating time the pack is temporarily disbanded. In March or April the female gives birth to her litter in a hole dug by both sexes. The father, sometimes with the help of other solitary adults, brings back food for the young wolves. In years of food shortage it is often only the highest ranking animals in the pack that pair. This functions as an efficient population control mechanism. At two to three months old the young wolves are first able to join the hunt themselves, though it is not until the following spring that they are sufficiently experienced to be independent. Although they can live up to 20 years in captivity, wild wolves seldom live for more than ten years. At this stage their teeth are very worn and they have lost a lot of their strength. Very old animals are often expelled from the pack and then live a solitary life until their death.

Normal posture (left) and hunting posture (right). Note lowered head and raised shoulder-hairs in the hunting posture.

Wolf howling (above)
When pack members meet they signal their relative status by the positions of their tails (below)

Golden Jackal

Canis aureus

(Dog Family)

Identification: HB 70-85 cm, T 22-27 cm, W 7-13 kg; resembles wolf (p. 154) but noticeably smaller; coat rather tawny, flecked with black; throat and underside whitish; black tipped tail; produces short barking noises as well as a long drawn out howl.

Distribution: Southeast Europe; southwest Asia; North and East Africa.

Habitat: Steppes and lowland scrub; often found close to habitation.

Behaviour: Active by day and night; usually hunts alone, more rarely in small packs; pairs for life.

Food: Wide range of small animals, carrion, berries, also attacks domesticated animals, eats fruit and scavenges in rubbish.

Breeding: Mating season Jan-Feb; gestation 9 weeks; 3-8 blind young with grey hair which open their eyes at 2 weeks and are independent at 6 months old.

Golden Jackals usually mate for life. In spring both sexes excavate a burrow in which the young will be born. Occasionally they will take over an existing burrow system, that of a fox or badger for example, and enlarge this to their own requirements. As with all dogs, the young are born fully furred but blind. The male helps to rear the young by bringing prey back to the den. The pups are suckled for two months but start to supplement their diet with meat as early as two weeks. At about 10 weeks old they start to practice hunting with their mother. By the autumn they are usually ready to hunt for themselves and start to live an independent life.

With the possible exception of the wolf, Golden Jackals have no notable natural enemies, although they do sometimes compete for food with other mammals such as Lynx (p. 170), Fox (p. 160) and sometimes Raccoons (p. 142). Although Golden Jackals often hunt alone they sometimes band together in small packs for hunting. Co-operative hunting enables them to tackle prey as large as sheep and goats. Often one jackal will chase the prey towards other members of the pack. Prey animals are seldom eaten where they are killed but are usually dragged to the safety of cover. Jackals are often quite bold in their behaviour and they sometimes attack domesticated animals. They also carry rabies. For these reasons they have long been persecuted by people. In spite of this they have been expanding toward the northwest in recent years and reached northeast Italy by 1989.

Jackals make a characteristic howling, often interspersed with barking noises, particularly in the evening

Golden Jackal: build is smaller and more delicate than wolf

Red Fox

Vulpes vulpes

(Dog Family)

Identification: HB 60-90 cm, T 35-50 cm, W 4-10 kg; head long and pointed; ears large, upright and triangular; tail bushy; colour very variable, from yellowish-red to black, but most commonly the characteristic fox red with white chin, belly and tip to tail; back of ears and feet black; makes a wide range of noises including a hollow bark, a scream and various kinds of chattering and whining.

Distribution: The whole of Europe (except Iceland); Asia, North Africa, North America. Widespread in Britain and Ireland.

Habitat: Inhabits virtually every kind of habitat, from high mountains to the coast, to steppe country, woodland and even urban environments.

Behaviour: Mainly active at dawn, dusk and at night; usually solitary outside the breeding season; occupies territories of up to 50 km^2; uses underground burrows when breeding, but usually spends the day lying amongst thick vegetation.

Food: Small mammals, birds, insects, carrion, berries, earthworms.

Breeding: Mating season Jan-Feb; gestation 52 days; 4-6 (maximum 12) cubs which are blind at first and have grey woolly fur; male helps to rear them by bringing prey to the den; young independent at 3-5 months.

Note: Foxes are a major carrier of rabies in continental Europe.

The Red Fox has long had a reputation for cunning and slyness, and not without good reason. It is an amazingly adaptable mammal and this, combined with its high reproductive rate, means that it has managed to overcome many barriers and has successfully colonised an increasing number of new habitats. Foxes are now even found in inner cities and suburban areas in many parts of their range. During the mating season in winter pairs of foxes can often be heard calling to each other at the start of the new breeding period. At this time the male fox often follows the female (vixen) for weeks at a time before mating. Towards the end of the gestation period, and also for a few days around the time of the birth, the vixen stays in the den and relies on food brought to her by the male. At about two weeks old the cubs open their eyes and their milk

Fox pouncing on mouse

teeth appear. At this stage they are able to start eating solid food, at first regurgitated and later on prey brought into the den. When they are one or two months old the young foxes begin to explore the surroundings of the den and they begin to develop the typical adult markings. While young, the cubs indulge in a lot of play that helps them to develop the skills they need in adult life. In late autumn they are left to fend for themselves and seek new territories. (Contiued p. 160).

Fox in mountain habitat (above)
Fox on the hunt (below)

Red Fox (continued)

Foxes indicate the borders of their territories by scent markings, using either urine or a secretion from the anal glands. They also have scent glands on their feet which leave a scent trail along regularly used tracks. This helps the foxes communicate with each other and enables them to follow their regular tracks during the night. Foxes also have an additional scent gland known as the violet gland on the top of the tail, the position of which can be seen by a small patch of darker fur. The precise function of this gland is not known but in males it is particularly active in the breeding season and is probably involved in communication between the sexes at this time of year.

Foxes also have an extremely well developed sense of hearing and can hear a mouse squeak at a distance of 100 m. They eat a large number of small rodents and rabbits which makes them useful predators in cultivated areas. Nevertheless their reputation for taking pheasants, hares and even occasionally young Roe Deer has led to foxes being regarded as vermin by gamekeepers and landowners. In reality, it is often only the unhealthy larger mammals that foxes are able to catch they can actually help to improve the overall health of such species. Where alternative prey is scarce foxes switch their diet to include insect larvae, crickets and earthworms or even plant food such as berries and fruits.

Foxes have been persecuted for centuries: either as vermin or for their thick winter coats, or more recently in continental Europe because they are an important vector of rabies. In just 50 years this disease has developed into an epidemic in much of Europe. The rabies virus is transmitted in the saliva of an infected animal, usually when it bites. Mice often carry the virus and may infect foxes by biting them in the lips when they are attacked. A rabid fox tends to bite very aggressively and thus readily transmits the disease further. Millions of foxes were shot or gassed in an attempt to halt the spread of rabies but such attempts have had no significant effect on the spread of the disease. More recently foxes have been fed an oral vaccine (via bait) and this is having some success in some areas such as Switzerland, Germany and parts of France.

In much of lowland Britain foxes are the subject of organised hunting by hounds; elsewhere they are generally persecuted for their devastations of livestock, especially lambs, although most studies have shown that fewer than 1% of lambs are lost to foxes.

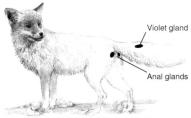

Violet gland

Anal glands

Positions of major scent glands

Fox showing characteristic face markings (above)
Cubs at play (below)

Arctic Fox

Alopex lagopus

(Dog Family)

Identification: HB 47-70 cm, T 30-50 cm, W 2.5-8 kg; similar in shape to Red Fox (p. 160) but with longer legs, shorter ears and also a somewhat shorter nose; there are two colour types: (a) grey-brown in summer, turning pure white in the winter; (b) brown or ash-grey throughout the year, with bluish tinge in the winter (this phase is also sometimes known as the Blue Fox and is rather rare, accounting for about 5% of the wild population); hoarse rather dog-like barks and yelps.

Distribution: Arctic and subarctic parts of both Eurasia and North America, including Iceland; has threatened status in Scandinavia and Finland.

Habitat: Open tundra north of the tree line; around coasts of the Arctic Ocean.

Behaviour: Active by day and night; not particularly shy; solitary outside the breeding season; digs extensive burrows; in summer territorial, but wanders in the winter.

Food: Small mammals, birds, eggs, fish, carrion, berries, fungi.

Breeding: Mating season Mar-Apr; pair for life; gestation 52 days; young (4-12 per litter) have dark fur and are independent at 3 months.

Few mammals are as well adapted as the Arctic Fox to the harsh climate of the far north. In the short Arctic summer, Arctic Foxes feed mainly on lemmings and voles, supplemented by berries and fungi. In coastal areas they eat mainly sea birds and dead animals washed up on the shoreline, occasionally they even eat seaweed.

In winter when the tundra is often covered by a deep layer of snow and ice, Arctic Foxes wander long distances (often hundreds of kilometres) across the ice fields in search of food. They often follow Polar Bears, scavenging on their kills. They may occasionally find small rodents by digging into the snow, particularly in areas where Reindeer herds have been feeding. Occasionally they will eat Reindeer droppings but in spite of their resourcefulness many Arctic Foxes do not survive harsh winters. Their remarkably high reproductive rate means that Arctic Fox populations can quickly recover from such losses. Their populations tend to show a three-year rhythm of increase and decrease, reflecting changes in the lemming numbers (see p. 66).

During the breeding season Arctic Foxes produce loud calls that can be heard over long distances. The foxes pair up, establish their territories and start to dig out their burrows. The birth of the young in May or June coincides with the period of maximum prey availability in the tundra. Both parents feed the cubs until the end of the summer, at which stage the young become independent and leave the family group.

Winter coat (above)
Transitional coat markings (centre)
Cub (below)

Raccoon Dog

Nyctereutes procyonoides

(Dog Family)

Identification: HB 60-80 cm, T 15-25 cm, W 5-10 kg; a rather squat, fox-sized animal with short legs; ears almost covered by body hair; long hairs on sides of head; bushy tail; coat brownish-grey to ochre with a black facial mask.

Distribution: Originally northeast Asia; introduced into the European parts of Russia and spreading to Finland and southeast and central Europe.

Habitat: Deciduous and mixed woodland with thick undergrowth, often by water.

Behaviour: Mainly nocturnal; solitary; swims well; lives in burrows in the earth or in natural holes; winter dormancy in the north of its range.

Food: Omnivore: rodents; birds, eggs, fish, frogs, insects, fruits, roots.

Breeding: Mating season Feb-Apr; gestation 7-8 weeks; litter size 6-7 (sometimes up to 12); the young have black fur and are blind at first.

The dark facial mask resembles that of the Raccoon but the Raccoon Dog has a uniform brown unringed tail.

Raccoon Dogs hunt slowly through thick undergrowth and their main prey is small mammals such as voles, although they also take a wide range of plant food as well. Unlike most members of the dog family, the Raccoon Dog does not flee on approach of danger but tends to freeze and remain motionless in the undergrowth, relying on its camouflaged coat for protection.

Raccoon Dogs will often take over abandoned Fox or Badger burrows but if these are not available they will dig a burrow system for themselves. In the north of its range the Raccoon Dog has a period of intermittent winter dormancy; this is unique for a member of the dog family. With their short legs, even a relatively shallow layer of snow presents problems for them.

Raccoon Dogs were first imported from eastern Asia into western Russia in about 1930. They subsequently escaped from fur farms but were also deliberately released to allow them to be trapped for fur. Since that time the species has continued to expand its range westwards, reaching central Europe in the early 1960s. The fur of the Raccoon Dog is thick and warm but relatively rough.

Raccoon Dog (left) compared with Raccoon (see p. 142)

Raccoon Dog in front of hollow tree (above)
Note short-legged appearance due partly to long hair (below)

Wild Cat

Felis silvestris

(Cat Family)

Identification: HB 50-80 cm, T 28-35 cm, W 5-8 kg; female somewhat smaller and lighter than male; resembles domestic tabby cat; fur grey to yellowish-brown with darker stripes; often white below; tail with black rings and rather bushy and blunt; eyes yellow-green with vertical slit-like pupils; nose flesh-coloured.

Distribution: Europe (except Scandinavia, Iceland and Ireland), Turkey, Caucasus; Africa. In Britain confined to the Highlands of Scotland.

Habitat: Mainly woodland and scrub.

Behaviour: Mostly nocturnal or active at dawn and dusk; solitary outside breeding season; territorial; marks boundaries with scent glands; mainly hunts on the ground but also climbs well; makes dens in hollow trees or rock crevices.

Food: Mainly mice and voles, also birds and small mammals up to about the size of a hare.

Breeding: Mating season Feb-Mar; gestation 63-68 days; litter size 2-6; the kittens have thin yellowish fur with darker spots; kittens open their eyes at about 10 days old, and become independent at 3-4 months.

As recently as 150 years ago Wild Cats were widespread in the forests of central and southern Europe and in Britain, but intense persecution has left just a few small pockets. However, Wild Cats are no longer considered a threat to game animals and are now protected.

Wild Cats mark their territories using scent glands. Scent marking involves the spraying of urine (which includes secretions from the anal glands) and rubbing on objects in the territory (there are also scent glands in the lips and cheeks). Wild Cats also have glands between their toes and they repeatedly scratch tree trunks, probably simultaneously sharpening their claws and leaving behind scent marks.

Like domestic cats, Wild Cats are very vocal in the breeding season and it is only at this time of year that the female tolerates the male in her territory. The kittens (usually two to four) are born in May and the den is usually in a hollow tree, although sometimes in an abandoned Fox or

Simultaneous claw sharpening
and territory marking

Badger burrow. The breeding chamber is often lined with grass, fur or bird feathers. The young start to leave the nest at four or five weeks old and rehearse their adult behaviour through play. At two months old they start to accompany their mother on hunting forays.

The integrity of the wild species is threatened by interbreeding with domestic cats. This interbreeding also results in fertile hybrids, although the domestic cat is almost certainly descended from the North African subspecies *Felis silvestris lybica*.

The Wild Cat's tail is bushy and rather blunt (above)
Wild Cats snarl when frightened (below)

Lynx

Lynx lynx

(Cat Family)

Identification: HB 80-110 cm, T 15-25 cm, W 18-35 kg; fur yellowish-grey to red-brown with variable darker (sometimes black) markings; noticeable beard on cheeks; ears with long black tufts; tail short with black tip; legs long; paws broad and hairy with retractable claws.

Distribution: Formerly throughout Europe but today restricted to Scandinavia and the Carpathians with isolated pockets in the Pyrenees, Rumania, Slovakia, Greece and the Balkans; central Asia; North America. Reintroduced to several parts of Europe including Switzerland, France, Slovenia and Austria. Probably became extinct in Britain long before the Roman period.

Habitat: Forests with rich undergrowth and clearings; also steep, rocky mountainous areas.

Behaviour: Active during the day and at night; solitary except in the breeding season; climbs well but rarely; regularly marks its territory with droppings; kills prey by stalking and ambush, not in long pursuit; rests in dense undergrowth or amongst rocks.

Food: Mammals ranging from mice to young deer, particularly Roe Deer and hares; occasionally birds.

Breeding: Mating season Feb-Mar; gestation about 10 weeks; litters of 2-4 kittens which open their eyes at 16 days, are suckled for 5 months and stay with their mother for about a year.

The Lynx was virtually eliminated from most of Europe in the nineteenth century and it is now extremely difficult to spot in the wild. However they have been successfully reintroduced in a number of regions in recent decades. Lynxes pose no danger to people and they also have little effect on wild populations of game animals. Although mainly nocturnal, Lynxes are sometimes active by day, particularly during the mating season or when feeding young. Their territories can be very large (20-150 km²) depending on prey density, and adjacent territories may overlap to some extent. Neighbouring animals tend to wander through their territories at different times and they leave behind scent marks on prominent places. Other Lynxes are able to 'read' these marks and thus avoid conflict.

Despite their relatively large size, Lynxes are classified as belonging to the small cats. Unlike lions and tigers they cannot growl but make purring noises similar to domestic cats. (Continued p. 172).

Note pointed ear tufts and bearded cheeks

Lynxes rest by day in the undergrowth (above)
Mother and kitten (below)

Lynx (continued)

Lynxes have very acute vision and hearing: a Lynx can spot a buzzard in the air at a distance of 4 km and can hear a twig break at 60 m. The long hairs on the sides of the face form a ruff which probably acts to concentrate sounds rather like the facial disc of an owl. They also have fully retractable claws.

When hunting, a Lynx watches carefully from a vantage point. When it spots a possible prey it creeps slowly towards the victim, approaching to about 20 m before

The fully retractable Lynx claw

catching the prey with an explosive spring and a final jump of up to 5 m. Prey animals which escape this attack are rarely followed for more than about 50 m. Like most cats, Lynxes do not chase their prey for long distances and they can only keep up a high speed for rather a short period of time.

Large animals such as Roe Deer or young deer are pulled to the ground and killed with a swift bite to the neck. The prey is then dragged away to a safe place before being eaten and remnants may be hidden under snow or leaves and consumed later. A Lynx needs about 1-1.5 kg of meat a day and can therefore survive for one or two weeks on a large kill.

Its wide paws give the Lynx mobility even over quite soft snow. Although it is much smaller than the Lynx the Wild Cat sinks much deeper into the snow because it has relatively much narrower paws. Because of this difference Wild Cats cannot normally survive in Lynx territory.

Size comparison between Lynx (behind) and Wild Cat (see p. 168)

The closely-related **Pardel Lynx** (*Lynx pardina*) is found scattered in montane forests of the Iberian Peninsula and also in the delta of the Guadalquivir River in southwest Spain. It is very similar to the Lynx, but somewhat smaller with a thinner coat and rather smaller spots. The Pardel Lynx is an endangered species; there are probably only about 100 living in the wild.

Lynx on the prowl (above)
Hunting in water (below)

Common Seal
Harbour Seal

Phoca vitulina

(True Seal Family)

Identification: HB 1.45-2 m, W 60-100 kg; female somewhat smaller and lighter than the male; coat pale grey to grey-brown with grey or black spots on the back; head rounded with short snout; external ears absent; hind flippers always directed backwards.

Distribution: Coastal waters of the north Atlantic Ocean and north Pacific Ocean; in Europe from Portugal to northern Scandinavia and Iceland, including Ireland, Scotland and parts of eastern England especially the Wash.

Habitat: Sandy coasts, estuaries, sheltered rocky coasts, sometimes freshwater near the coast.

Behaviour: Diurnal; gathers into groups during the breeding season but otherwise solitary; tends to remain faithful to particular sites; swims and dives well and for long periods; very clumsy on land; often rests and sunbathes on sand banks.

Food: Fish, crustaceans, squid and whelks.

Breeding: Mating season June-July; gestation 11 months (includes delayed implantation); normally a single pup (rarely twins) which can swim within hours of being born, is suckled for 8 weeks and is then independent.

Note: Many European colonies have declined recently.

Like all seals this species lives an amphibian life. Seals use their limbs as paddles and they can move through the water at great speed, aided by their hydrodynamic profile. When swimming the main thrust is provided by the flexible body moving the hind flippers from side to side. However, seals are very ungainly when on land. In the true seals, as opposed to the eared seals or sealions, the hind limbs are virtually useless on land. Forward movement on land is by heaving the body forwards on the belly.

When underwater, seals can close their nostrils and ear openings to prevent the water from entering. When diving after prey Common Seals usually stay under water for five to ten minutes, although they can remain submerged for as long as 45 minutes in an emergency. Their blood contains a high concentration of red corpuscles which bind oxygen very efficiently. They are also able to limit their oxygen requirements when underwater. Adaptations to this submarine ability include the heart rate reducing from about 150 to 10-15 beats a minute and the blood supply diverting away from the skin, muscles and intestine, such that only the heart and brain continue to receive the normal blood flow. (Continued p. 176).

Humping forwards on land

Resting on sandbank (above)
Resting on rocks (below)

Common Seal (continued)

Common Seals eat mainly fish which they hunt during the day in shallow coastal waters. They use their sharp powerful teeth to catch and eat the fish underwater. Each seal can consume about 5 kg a day and this explains why seals are often regarded as competitors by fishermen. In addition, Common Seal skins are also used in the fashion industry. Moreover, the remaining colonies of Common Seals are now under increasing pressure from marine pollution, particularly in the North Sea and Baltic Sea.

Common Seals often give birth to their pups on sand banks or shelving coasts which are exposed only at low tide. To cope with this situation the newborn pups are able to swim when they are only a few hours old. Common Seal pups usually lose their foetal white woolly coat just before they are born and the fur of the newborn pups therefore usually closely resembles that of the adult. This coat is fully waterproof, another adaptation that enables the young pup to swim almost immediately.

For the first few weeks after being born the pup will stay close to the mother. The pup will be suckled both in the water and on land. Occasionally Common Seals produce twins, but the mother can usually manage to look after only a single pup. This occasionally results in an abandoned pup being left alone on a sandbank or shoreline. Such orphans often die unless they are rescued and cared for carefully in special seal sanctuaries.

Common Seal diving

Common Seal adult (above)
Common Seal pup (below)

Grey Seal

Halichoerus grypus

(True Seal Family)

Identification: HB 1.8-3.3 m (female up to 2.2 m), W 120-300 kg (female up to 200 kg); grey to brownish, male with pale patches on a dark background, female with dark patches on pale background; head with elongated and rather heavy muzzle; often has small external ear visible.

Distribution: Atlantic coasts of central and northern Europe; North Sea and Baltic Sea, east coast of Canada. Around most British and Irish coasts except for those of the English Channel.

Habitat: Rocky coasts, sometimes on sandbanks.

Behaviour: Diurnal; tendency to remain faithful to particular sites; rarely comes onto the land except during breeding season and moult; dives very well (often for 20 minutes at a time and to depths of over 100 m); in breeding season a male accompanies a harem of up to 10 females.

Food: Fish, crustaceans, molluscs and occasionally birds.

Breeding: Mating season Oct-Nov in the Atlantic Ocean, Feb-Mar in the Baltic Sea; gestation 50 weeks (includes delayed implantation); the single pup is born with a white foetal coat (lanugo) which is shed after 2-3 weeks; it is suckled for 2-3 weeks and learns to swim at about 4-5 weeks old.

Unlike Common Seals, Grey Seals tend to prefer rocky coasts and often remain at sea for months at a time. Grey Seal populations have been heavily persecuted in the past and their range and numbers are much reduced. The largest European populations are now in the British Isles; there are 45,000 adults breeding in the Outer Hebrides for example.

Newborn Grey Seals have a white crumpled protective coat which they shed after two or three weeks. The pups are left on land and their mothers return to suckle them about twice a day. Seal milk contains about 50% fat and is five times as nourishing as cows' milk. By the time the pups have moulted into their adult coats they have roughly trebled their birthweight and at this stage they become independent of their mothers. The pups then make their own way to the sea and start to forage for themselves. Such inexperienced young seals have a pretty low success rate when hunting and year old seals often weigh less than they did when weaned. Some 60% of young Grey Seals die before they are a year old.

The bulls come ashore at about the same time as the cows and defend their territories, each of which will include a small group of suitable cows. Mating takes place about two weeks after the pups are born.

Profiles of Grey Seal (left)
and Common Seal (see p. 174)

Grey Seal cow (above)
Grey Seal bull (below)

Ringed Seal

Phoca hispida

(True Seal Family)

Identification: HB 1 2-1,5 m, W 45-100 kg; female somewhat smaller and lighter than male; resembles Common Seal (p. 174) but smaller; coat ochre to grey-brown with rather irregular darker patches, each with a pale border; yellowish-white below.

Distribution: Baltic Sea, some inland lakes in Finland and northwest Russia; Arctic coasts of North America and Eurasia.

Habitat: Coastal waters, fjords and amongst sea ice; often enters fresh water.

Behaviour: Solitary; adults remain faithful to particular sites; often rest on the ice, rarely on land; can dive to 300 m and for up to 20 minutes.

Food: Small crustaceans, fish.

Breeding: Mating season Apr-May; gestation 11 months (includes delayed implantation); a single white-coated pup per litter, usually born in a snow cave; pup suckled for 4 weeks, after which it moults and becomes independent.

Ringed Seals create breathing holes in the ice when it is quite thin in the autumn and maintain these by constant use as the ice thickens through the winter. They have a thick layer of blubber below the skin which serves both as a food store and also to insulate them from the intense cold of their environment. In extreme weather they shelter in holes dug into the snow. The young are usually also born in snow caves (lairs) which may be 4-5 m long. Pups are usually born in March or April.

Harp Seal

Phoca groenlandica

(True Seal Family)

Identification: HB 1.55-2.2 m (female to 1.85 m), W 120-220 kg (female to 180 kg); similar to Common Seal (p. 174) and Ringed Seal but larger; coat grey to yellowish-brown, silver-grey beneath; adults have saddle-like dark markings on the back; in the male this is roughly harp-shaped, hence the name, in the female the markings are usually paler and broken up into spots.

Distribution: North Atlantic Ocean, Arctic Ocean; breeding grounds in the White Sea and around Greenland, Newfoundland and northern Iceland.

Habitat: Sea ice and open sea.

Behaviour: Sociable, in small groups or large herds; travels long distances around the ice; dives up to 300 m and for 30 minutes at a time; keeps open breathing holes in the ice.

Food: Fish, crustaceans.

Breeding: Mating season Mar-Apr; gestation about 11 months (includes delayed implantation, see p. 251); a single white-coated pup per litter, born on the ice and suckled for 3-4 weeks.

Harp Seals spend almost their entire lives at sea and even the pups are born on floating pack ice. The pups are born with a soft, silky white coat and many are killed for the fashion industry. The pups begin to lose this soft coat after a week and so are only valuable for their pelts in the first two or three days of their lives.

Two other Arctic seals occur on the north coasts of Scandinavia and Iceland: the **Bearded Seal** (*Erignathus barbatus*) is uniform brown with very prominent whiskers; the **Hooded Seal** (*Cystophora cristata*) is pale grey with irregular dark patches and, in the adult male only, an inflatable proboscis.

Ringed Seal (above)
Harp Seal with newborn pup (below)

Mediterranean Monk Seal

Monachus monachus

(True Seal Family)

Identification: HB 2.3-2.8 m, W 180-320 kg (male to 250 kg); coat brown or blackish-brown above, yellowish-white underneath, external ears absent; rear flippers always directed backwards; the front flippers with powerful claws.

Distribution: Mediterranean and Black Seas, Mauritania, Atlantic coast of Morocco, Madeira and Canary Islands.

Habitat: Undisturbed sandy or rocky coasts, isolated bays.

Behaviour: Diurnal, lives in small colonies; sticks to traditional breeding sites; swims and dives well; also able to climb over rocks well; not particularly territorial even in the breeding season.

Food: Fish, squid, crustaceans.

Breeding: Mating season Oct-Nov; gestation about 11 months (includes delayed implantation, see p. 251); a single pup born every 2 or 3 years; pup has white coat at first, is suckled for about 5 weeks and remains with its mother for 2-3 years.

Note: This species is highly endangered.

Monk Seals stick to particular sites throughout the year and this, combined with their tameness, makes them particularly vulnerable to human disturbance. Whereas once there were tens of thousands in the Mediterranean, only about 500 remain, scattered in small groups, particularly in remoter parts of Greece and the Atlantic coast of Mauritania.

Even though they live in relatively warm conditions, Monk Seals have a thick layer of fat under the skin just like their Arctic relatives. On hot summer days they tend to seek out caves and the shadows of cliffs and move very little. Bulls do not stake out territories in the breeding season but follow individual cows in the water before mating.

The pups, which are born on land, can swim from an early age and stay close to their mother. They stay with their mothers for up to two or three years and only then will she become pregnant again.

There is a closely related species of Monk Seal in Hawaii; another that formerly occured in the Caribbean is now extinct.

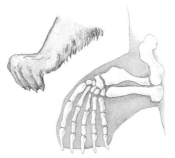

Front limb of Monk Seal and skeleton. Monk seals use their claws to clamber up even quite steep rocky coasts

Like all seals, Monk Seals can close their nostrils before diving

Walrus

Odobenus rosmarus

(Walrus Family)

Identification: HB male 3.1-4.5 m, female 2.8-3.4 m, W male 1000-1500 kg, female to 800 kg; young have thick reddish-brown coat, adults yellowish-brown and getting increasingly naked as they grow older. Skin thick and wrinkled; upper lip has stiff bristles and the upper canines are elongated as tusks; external ears absent.

Distribution: Coasts of the Arctic Ocean and north Atlantic Ocean, Spitzbergen, Iceland, northern Norway and Russia. Vagrants occasionally reach Scotland.

Habitat: Shallow coasts and coastal ice, in winter moving to relatively ice-free areas.

Behaviour: Lives in family groups (consisting of a male, 2-3 females and several young) and also in male-only herds; very clumsy on land but excellent swimmers, often over long distances; dives for only about 5-10 minutes up to a depth of 30 m; uses tusks to dig into the seabed when seeking food.

Food: Molluscs, worms, fish and occasionally carrion and seals.

Breeding: Mating season Apr-June; gestation about 12 months; a single pup born every 2 years; pups can swim from a very early age and are suckled for the first 18 months to 2 years.

The Walrus and the Elephant Seal of the South Pacific are the largest of all the seals. The whiskered upper lips and long tusks make the Walrus unmistakable. They use their tusks to grub about in the sea bed for small marine invertebrates such as bivalve molluscs, crustaceans, echinoderms and small fish. Most of their other teeth fall out at an early age. Walruses are not such active predators as the other seals and they rarely take large prey.

The young are born in April or May, usually in shallow coastal waters. Pups are about a metre long and weigh around 30 kg at birth. Walrus mothers are very protective of their young and accompany them everywhere at first. They sometimes even give the pups rides if they get tired from swimming.

Walrus populations have diminished markedly in recent years and the species is now considered vulnerable. They are hunted for their meat, blubber, leather (the skin is 2-3 cm thick) as well as for their tusks. The latter can weight up to 3 kg and are a source of valuable ivory.

Walruses sometimes use their tusks in territorial disputes

Walruses hauled out on a rock. Note the use
of hind limbs to help in locomotion (above)
The upper canines are modified as powerful tusks (below)

Przewalski's Horse
Wild Horse

Equus przewalskii

(Horse Family)

Identification: HB 2.2-2.8 m, S 1.2-1.45 m, T 90-110 cm, W 200-350 kg; coat yellowish-brown with a darker mane and stripe along back; muzzle and belly white; lower legs dark brown to black; the mane is dark brown or black, and erect, not drooping forward like the mane of a domestic horse; tail black.

Distribution: Originally found from western Europe to central Asia, but now probably extinct in the wild (last known from Mongolia and western China).

Habitat: Grass and forest steppe, semi-desert.

Behaviour: Active by day; capable of fast, sustained running.

Food: Grass, herbs.

Breeding: Mating season usually in the spring; gestation about 340 days; a single foal which can stand as soon as it is born and is suckled for about 6 months.

Note: This species is the ancestor of domestic horses.

At one time this original wild horse had a wide distribution from western Europe right across to central Asia and existed in three different subspecies: a southern Russian form (driven to extinction in the nineteenth century), a somewhat smaller forest form (also now extinct) and the eastern steppe form (Przewalski's Horse). The forest form inhabited central European woodland and became extinct early in the last century.

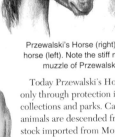

Wild horses live in small groups made up of several mares with their foals and led by a single stallion. The dominant stallion drives out all the young males from its herd and also protects his harem from rival stallions.

Przewalski's Horse (right) and domestic horse (left). Note the stiff mane and white muzzle of Przewalski's Horse.

Today Przewalski's Horse persists only through protection in animal collections and parks. Captive animals are descended from wild stock imported from Mongolia in the late nineteenth century. A herd has recently been established in southern France with the intention of returning them to Mongolia.

Tails of domestic horse (left) and Przewalski's Horse (right). Note that the tail of Przewalski's Horse has short hairs towards the base.

Adult grazing (above)
Mare feeding foal (below)

Wild Boar

Sus scrofa

(Pig Family)

Identification: HB 1.1-1.8 m, T 20-30 cm, W 50-180 kg (maximum 250 kg); male larger and heavier than female; stocky body and wedge-shaped head; dark bristly coat, in winter with thick underfur; dark brown to ochre, piglets pale with dark longitudinal stripes; ears upright; adult males have long teeth forming short tusks.

Distribution: Most of Europe except Scandinavia, Iceland and British Isles (extinct in the British Isles since the seventeenth century); central and southern Asia; North Africa; introduced to North and South America and Australia.

Habitat: Deciduous and mixed forests, meadows and marshes, up to the tree-line in mountains, coniferous plantations.

Behaviour: Active by day and night; lives in family groups; adult males solitary except in breeding season; establishes definite territories; fond of wallowing.

Food: Omnivorous: roots, bulbs, acorns and beechmast, fruits, herbs, also worms, insects, birds' eggs and small mammals as well as carrion.

Breeding: Mating season Nov-Jan; gestation 4-5 months; 4-6 (maximum 12) piglets which leave the nest after a week and are suckled for about 3 months.

Note: This species is the ancestor of most breeds of domestic pig.

Wild Boar are widespread in the forests of Europe and Asia and this species has also been introduced to America for sport. The size and weight of individuals increases towards the colder regions following a general ecological rule. A Wild Boar from the Carpathians for example can be up to five times heavier than one from Sardinia.

Wild Boar live in groups of between six and ten individuals consisting of females and young of different ages. Fully grown boars tend to live a solitary life. In summer, Wild Boars tend to rest during the day and set out on hunting forays in the evening, retiring again before dawn. In winter, on the other hand, they can more often be seen during the day and tend to spend the cold nights sleeping close together in a well protected hollow. At this time of year their thick underfur gives them good protection from the cold. Wild Boars tend not to be very active in deep snow as their heavy bodies easily sink into snow drifts. Under such conditions they tend to keep to regularly used tracks. (Continued p. 190).

Head of Wild Boar showing tusks

Each piglet has its own nipple which it defends vigorously from its siblings (above)
Wild Boar in winter coat (below)

Wild Boar (continued)

The Wild Boar mating season lasts from November until the beginning of February. Sexually mature males give off a penetrating scent at this time of year. During this rutting season the boars can inflict quite serious wounds on each other with their tusks. Their vital organs are quite well protected by a thick layer of cartilage over the thorax which acts as a shield during such fights. The strongest boars then mate with the females and the young are born four to five months later in a nest constructed from twigs and moss.

There are usually between four and six piglets per litter but young sows tend to have fewer and sometimes as many as 12 are born. At first the piglets stay in the nest and lie close to each other when the mother is away. Sometimes she covers the litter with nest material. After a week the piglets are ready to accompany their mother on foraging trips. At this time of year the sow is extremely protective of her litter and is likely to attack any intruder, including humans.

Piglet sucking milk from its mother

When danger threatens, the piglets scatter and lie low in the undergrowth using their camouflaged colouring to great effect. After about two or three weeks the sow will rejoin the herd. The sow continues to suckle her litter until they are about three months old, after which they forage for their own food. At this stage the piglets gradually lose their stripy colouring and show the dark bristly adult coat.

Many young Wild Boars die as a result of cold, wet weather, predation or from infestation with parasitic worms. In really bad weather only about 20% survive until the following spring. On the other hand, in certain parts of central Europe Wild Boar numbers have risen steadily, possibly because of the absence of natural predators such as Wolf and Lynx and reduced hunting pressure. Wild Boar root about in the ground when searching for food and can sometimes cause considerable damage to crops.

Wild Boar love to wallow and in hot weather
they will even dive into the water (above)
Small group of sows with piglets (below)

Red Deer

Cervus elaphus

(Deer Family)

Identification: HB 1.65-2.5 m, T 12-15 cm, W 100-340 kg; male (stag) markedly larger and heavier than female (hind); coat red-brown in summer, grey-brown in winter; rump patch yellowish; young calves spotted; adult males have branching antlers that are shed and regrown each season; in winter adult males have a well-developed mane; males make loud roaring noises during the rutting season.

Distribution: Whole of Europe except Iceland, northern Scandinavia, Balearic Islands, Sicily, Peloponnese and Crete; introduced to South America, Australia and New Zealand. Abundant in the Scottish highlands; scattered populations in England and Ireland.

Habitat: Deciduous and mixed woodland; montane forests up to and beyond the tree-line; parkland; moorland; river valleys.

Behaviour: Mainly active at dawn, dusk and by night; runs well and also jumps and swims; likes to wallow; females live together with young animals in herds; outside the breeding season the males roam about in small groups or are solitary.

Food: Grass, herbs, leaves, tree shoots, bark, fruit.

Breeding: Mating season Sep-Oct; gestation about 8 months; normally a single calf (rarely twins) which can run after a few hours, is suckled for 8-9 months and stays with its mother for about a year.

The most characteristic feature of the adult male Red Deer is its impressive set of antlers, which can weigh as much as 15 kg. Unlike horns, antlers are shed each year and replaced by a new set. The new antlers are already visible within a few weeks of shedding the old pair and they are at first covered by a soft layer of skin known as velvet. This skin carries nerves and blood vessels to nourish the growing antlers which are still soft at this stage, and the deer will avoid using its antlers as they are still quite sensitive. Disputes between stags in velvet involve kicking out with their front limbs. The new antlers grow quite rapidly – up to 6 cm a day. Antler growth is very expensive in energy terms and the stags may lose quite a lot of their body reserves at this season. The size and shape of the antlers depends partly on the age of the stag, but also on its state of nourishment and health. The state of the stag's antlers also affect its social standing. (Continued p. 194).

Dispute between two stags in velvet

Rutting Red Deer stags

Red Deer (continued)

The number of branches or points (known as tines) on the antlers is not a good indicator of the animal's age. Antler development appears to depend mostly on genetics and also on the food quality offered by the habitat. Really well-developed antlers may have as many as 12-14 branches or tines, and sometimes even more. In July the developing antlers become hard and the velvet skin starts to dry up and peel away in blood-stained tatters. At this stage the deer spend quite a lot of time rubbing their antlers against tree trunks and branches.

When the days begin to shorten in September the rutting season begins. The stags start to draw attention to themselves by roaring. Each mature stag seeks to gather as many as hinds together as possible and this often leads to disputes with other stags. Rutting stags charge towards each other, interlocking their antlers and pushing. Although these fights are quite energetic they seldom lead to serious injury. Each clash ends when the weaker stag withdraws.

The Red Deer hinds live together with calves in small herds, in which one adult female is usually dominant. In early summer the pregnant females temporarily abandon the herd while the young are born. Although the newborn calves are quite capable of standing and walking, they spend the first few days crouched on the ground hidden under the scrub or tall grasses waiting for their mother to come back and feed them.

In front of each eye lies a scent gland (termed the pre-orbital gland). This is a fold of skin which produces a particular scent important in communication between hind and calf. Soon after a new calf is born its mother learns to discriminate its own particular scent and she can home in on this scent from quite a long distance downwind. When danger threatens the calf crouches close to the ground and closes its scent glands. Hinds can even detect this interruption of the scent signal and will quickly return to the calf.

Position of scent gland in Red Deer.
Below right: scent gland when open

In Scotland, where most Red Deer live on open hill country, excessive numbers are causing serious damage to natural vegatation and preventing the regeneration of woodland. Red Deer are increasingly being domesticated and farmed for the production of meat.

Young calf sitting tight (above)
Red Deer hind with calf (below)

Fallow Deer

Dama dama

(Deer Family)

Identification: HB 1.3-1.7 m, T 15-24 cm, W 35-100 kg; male (buck) larger and heavier than female (doe); in summer coat red-brown with white spots, in winter grey-brown with less distinct spotting; rump patch white with dark borders; tail much longer than in other deer, with dark upper side; antlers flattened in older bucks; antlers shed Apr-May; fresh antlers complete their growth in Aug.

Distribution: Most of central Europe; absent from most of Scandinavia and from Iceland; formerly more widespread but now only pockets in southern Europe; introduced to North and South America and New Zealand. Widespread in England and Wales, more scattered in Scotland and Ireland.

Habitat: Deciduous and mixed woodland in lowland areas, open parkland.

Behaviour: Predominantly crepuscular; females live together with calves in large herds; males in smaller groups; males rut over females in the autumn.

Food: Grasses, herbs, fruit; in winter also bark.

Breeding: Mating season Oct-Nov; gestation about 32 weeks; litter usually a single fawn (occasionally twins) which lies for 2-3 weeks in the undergrowth before joining the herd; fawn suckled for 6-9 months and stays with mother for a year.

Fallow Deer were spread to many parts of southern Europe, first by the Phoenicians and later by the Romans, mainly from their stronghold in Turkey and some other pockets in the Mediterranean region. Since the Middle Ages this species has been spread further through deliberate introductions to parks and forests over much of central and western Europe including the British Isles, and artificial selection has led to the existence of a number of colour forms. These include almost completely black deer, pale brown, and almost white individuals. These unusual colour varieties were highly prized in royal parks and have therefore persisted. Although domesticated Fallow Deer herds are remarkably tame, wild herds are very shy.

Their senses of hearing and smell are as acute as those of the Red Deer, but they can see even better and unlike most other deer species they can also differentiate between stationary objects. This makes them particularly difficult to hunt. In the early autumn the bucks disperse from their small herds and begin to rut in disputes over adult females. Rival males can gauge each other's strength by the size of their antlers. Fights therefore only usually take place between stags of similar stature. Rutting Fallow Deer bucks produce a distinctive noise rather like a belch or groan.

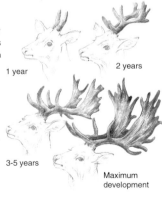

1 year
2 years
3-5 years
Maximum development

Development of antlers with increasing age

Two young bucks in velvet (above)
Doe with fawn (below)

Sika Deer

Cervus nippon

(Deer family)

Identification: HB 1-1.5 m, T 14-19 cm, W 60-130 kg; smaller than Red Deer; coat red-brown in summer, speckled yellow-white at all ages; grey-brown in winter, unspotted or spots indistinct; antlers with at most 3 or 4 tines; makes penetrating whistling sounds in rutting season.

Distribution: Introduced from Japan and northeast China to many sites in central, western and eastern Europe; also to Australia, New Zealand, South America and Madagascar. Scattered populations throughout England, Scotland and Ireland.

Habitat: Deciduous and mixed woodland; parks.

Behaviour: Active by day and night; females live together with young animals in herds; males live in small groups or are solitary.

Food: Grass, herbs, tree shoots, fruit, berries.

Breeding: Mating season Oct-Nov; gestation 32 weeks; usually a single calf (rarely twins); development as for Fallow Deer (p. 196).

This rather small, elegant deer from the Far East has now been introduced to Europe, mainly to parts of the British Isles, France, Denmark and Germany.

Sika Deer are not particularly prized by huntsmen, mainly because their antlers are not as impressive as those of Red Deer. There is a problem with interbreeding in some areas, threatening the integrity of both species.

The delicate hooves of Sika Deer are unsuitable for travel over snow, and some die from exhaustion in snowy winters, even though they are relatively insensitive to cold. They can swim well and leap high fences and other barriers, but they cannot run for long periods and sometimes fall prey to feral dogs. When danger threatens, Sika Deer signal to the rest of their herd by flashing their white rump patches. This is done by erection of the surrounding dark hair.

American White-tailed Deer (*Odocoileus virginiarus*) have been introduced to southern Finland. They are between Red and Sika Deer in size, a uniform brown, with a large tail which is raised to display white underside and rump when alarmed.

The spreading white rump patch acts as a signal to other members of the herd

The white spots are particularly obvious in summer (above)
The new antlers begin to bud out just a few weeks after the old antlers are shed (below)

Axis Deer

Cervus axis

(Deer Family)

Identification: HB 1.1-1.4 m, T 20-30 cm, W 75-100 kg; coat red-brown
with white spots and a white throat; belly and inner sides of legs also white;
tail long; male usually has 6 tines to antlers.

Distribution: Native to India and established as isolated introductions in
Europe: Northeast Italy, Istria, Poland, Ukraine; also introduced to South
America, Australia and New Zealand.

Habitat: Deciduous woodland, river valleys, park landscapes, particularly
close to water.

Behaviour: Crepuscular and also active by day; swims well; often takes
refuge in the water; lives in large mixed herds; older males solitary.

Food: Grasses, herbs, foliage.

Breeding: Mating season variable, may be at any time of year; gestation at
least 7 months; 1-2 calves; development as for Fallow Deer (p. 196).

Axis Deer live naturally in dense forests in India. Animals
of all ages and at all seasons have clear white
spots. There have been many attempts to
introduce this attractive deer to Europe, but
few have been successful largely because it is
very susceptible to cold.

As with many tropical deer, the mating
season may be at any time of year and
individuals may enter the rut in almost any
month. Consequently, calves may also be
born at any time of the year. Likewise
shedding and regrowth of antlers can take
place at any time.

Adult Axis Deer showing six points on antlers

Chinese Muntjac

Muntiacus reevesi

(Deer Family)

Identification: HB 90-100 cm, T 15 cm, W 9-18 kg; much smaller than
other deer; uniform brown, no white on rump with tail down, but tail raised
when alarmed showing prominent white underside. Antlers small,
single-pointed, on prominent permanent pedicels.

Distribution: Native to China; introduced and widespread in much of
England, Wales and parts of Scotland.

Habitat: Woodland with shrub layer.

Behaviour: Solitary; active day and night. Makes a loud bark, often repeated
many times.

Food: Mainly leaves of shrubs such as bramble, rasp, hawthorn and ivy.

Breeding: Throughout the year. Gestation 7 months, a single young.

Chinese Muntjac were first introduced in Bedfordshire in the early
twentieth century and by natural spread and additional introductions have
colonised most of England and beyond.

The **Chinese Water Deer** (*Hydropotes inermis*) is another introduced
species from China, now present in eastern England (Cambridgeshire,
Norfolk) and parts of France. It has no antlers but very large ears, and the
males have prominent tusks.

Axis Deer buck (above)
Chinese Muntjac (below)

Roe Deer

Capreolus capreolus

(Deer Family)

Identification: HB 1-1.4 m, T 2-3 cm, W 15-35 kg; male larger and heavier than female; coat red-brown in summer, grey-brown in winter; rump patch white; tail virtually absent, but in winter female has a tuft of white hair resembling a tail. Male antlers 6-pointed when grown; antlers shed in October; regrowth begins in November and the velvet is shed in April or May; male has a raspy courtship call and both sexes make a barking sound when alarmed.

Distribution: Whole of Europe except Ireland, Iceland, northern Scandinavia and the Mediterranean islands; northern Asia as far east as China.

Habitat: Mixed and deciduous woodland with thick undergrowth and clearings; woodland edges, river valleys, fields, parks, in mountains up to the tree-line.

Behaviour: Mainly active at dawn and dusk; runs well but not for long periods; jumps and swims well; in winter lives in loose groups of 3-30 animals, in summer solitary; marks territory with scent.

Food: Leaves, buds, herbs, fruits.

Breeding: Mating season July-Aug; gestation 7 months (includes delayed implantation, see p. 251); 1-2 young which lie under cover for the first week then follow their mother; fawns suckled for 2-3 months and independent at a year old.

In different parts of their range Roe Deer occupy a wide variety of habitats, from forested steppe through marshland, right up to high mountain terrain. Such adaptability has resulted in very large populations of Roe Deer in many areas, in spite of the fact that they are regularly shot as game and are not infrequent casualties of road traffic. Unusual colour varieties including white and black are not that uncommon.

The antlers of a fully grown buck typically have six points. As with other deer species these are antlers as opposed to horns; they are made of bone and regrown each season under a layer of soft skin or velvet. In April or May the velvet begins to peel off the fresh antlers and at this time of year the deer repeatedly rub their antlers against twigs and branches. The new antlers are then ready for the rutting season. At the end of the mating season the level of the sex hormone testosterone decreases in the Roe buck's body and this triggers the shedding of the antlers. (Continued p. 204).

Jan/Feb   March

April (loss of velvet)   Oct/Nov (fall of antlers)

Antler development through the year

Roe Deer doe (above)
Roe Deer often produce twins (below)

Roe Deer (continued)

Roe Deer have a characteristic almost round rump patch which includes the very short, almost invisible tail. In summer the rump patch is usually yellowish, but in winter it becomes bright white and somewhat larger. The rump patch works as a signal, helping to keep the herd members together, particularly in winter when the herds are larger. The herds begin to disintegrate in spring and at this time of year bucks and does tend to live a much more solitary life. Each buck marks the borders of its territory by rubbing a secretion from glands on the forehead against tree trunks and branches.

In the mating season bucks follow the does for long periods before mating. The fertilised egg then remains in a state of suspended development until the following spring. Interestingly, the Roe Deer is the only ungulate species for which this phenomenon has been recorded. The fawns are born in May or early June. Roe Deer usually produce twins.

For the first few days the mother leaves the fawns in a safe site, returning only from time to time to feed them. Not only do the young fawns remain absolutely still, they also produce no detectable scent – they can even remain undetected by hounds passing as close as within two or three metres. If a newborn Roe Deer is touched or lifted up, there is always the danger of transferring human scent to it and this may cause the mother to desert it. One should therefore never interfere with newborn Roe Deer fawns – the likelihood is that their mother is waiting nearby and will return.

Although Roe bucks usually lose their antlers in the winter, the sexes can still be distinguished by the shape of the rump patch. Only the doe has this tail-like tuft of hair at the bottom of the patch

Roe bucks can detect by smell whether a doe is ready to mate (above)
Adult Roe buck (below)

Elk

Alces alces

(Deer Family)

Identification: HB 2-2.9 m, S 1.5-2.1 m, T 5-10 cm, W 300-600 kg; largest living deer; male (bull) usually has flattened, palmate antlers which spread almost horizontally; female (cow) lacks antlers; head long; upper lip wide and overlapping mouth; male has beard; calf lacks spots.

Distribution: Scandinavia, northeast Europe, northern Asia; Canada, Alaska (where known as Moose).

Habitat: Deciduous and mixed woodland, particularly close to water or in swampy sites.

Behaviour: Active by day and night; solitary; covers large distances and swims well.

Food: Plants, leaves, shoots, water plants; in winter twigs, needles and bark.

Breeding: Mating season Sep-Oct; gestation around 36 weeks; 1-2 calves which follow their mother soon after they are born, are suckled for 3-4 months and stay with her for at least a year.

Note: The horse-sized long-legged Elk is the largest European wild animal.

The long drooping snout and massive antlers make the Elk unmistakable. The characteristic shovel-shaped antlers only develop in older males (those over five years old) and some have narrower antlers throughout their lives. As in all deer, the antlers are shed after the rut and are regrown each spring within two or three months.

Unlike most other large deer species the female Elks do not form separate herds but stay with their own families, sometimes accompanied by their young of previous years. In the autumn adult males often indulge in fierce fights. Males do not defend harems, but each male remains with a single female at a time, often leaving the first after mating to go in search of another.

Elks have wide, rather flat feet which help them keep a grip on marshy ground or over snow. Like Roe Deer they are very selective when feeding, tending to avoid poor grasses in favour of more nutritious food such as buds and shoots. In summer they often eat water plants, wading deeply to reach them.

Elks are often quite tameable and were used in the past to pull carriages or carry loads.

Elks often bend down on their knees when grazing

Bull Elk in velvet (above)
Cow with calf (below)

Reindeer
Rangifer tarandus

(Deer Family)

Identification: HB 1.8-2.2 m, T 10-20 cm, W 80-220 kg; female somewhat smaller and lighter than male; both sexes have antlers, but those of the female are smaller; summer coat brown, winter coat thicker and pale brown to whitish, often with pale grey mane; calves lack spots.

Distribution: Arctic Europe, Asia and North America.

Habitat: Open tundra, wooded tundra and in mountains to 2500 m.

Behaviour: Active by day; lives mostly in small herds; in winter larger herds of up to 1000 animals gather; undertake long yearly migrations; in rutting season males gather harems of up to 20 females; old males often solitary.

Food: Grass, herbs, shoots, leaves, lichens, fungi.

Breeding: Mating season Sep-Nov; gestation 7-8 months; usually a single calf, which can walk from the first day and stays with its mother for at least a year.

Reindeer undertake regular annual migrations between summer and winter quarters. They leave the forests in spring and move out on to the treeless tundra, covering distances of up to 1000 km. Small groups of Reindeer emerge from the woods and coalesce into enormous moving herds. It is during this journey that the calves are born, and they can walk and follow their mother within an hour or so of birth. When the herds arrive at their grazing grounds in the tundra, they split up again into smaller groups. The animals, which have starved somewhat during the winter, now fatten up quickly. Reindeer are highly selective feeders, and take only very nutritious food such as fresh shoots, buds, flowers and leaves. They tend to retreat to windy plateaux or towards the Arctic coasts, partly to avoid the attention of the millions of bloodsucking insects. At the end of the short Arctic summer the Reindeer assemble again for the return migration towards the forests where they can gain some protection from winter frost and storms. Here however it is much more difficult for them to find

nourishing food. They dig through even quite deep snow with their front hooves in order to get at the vegetation below. Reindeer are even said to be able to smell certain lichens such as so-called Reindeer Moss (*Cladonia rangiferina*) even when it is covered by a metre of snow.

Reindeer can spread their hooves, which then act rather like snow shoes (left and centre); compare the hoof print of Fallow Deer (right)

Reindeer are still herded in many parts of their range by people such as the Lapps and various Siberian traditional societies. They use the semi-domesticated Reindeer to provide meat, leather and milk. Such domesticated animals are often more varied in colour than the wild type and include white, dark brown and very variegated patterns.

Reindeer in velvet (above)
Bull Reindeer often have asymmetric antlers, a single flattened branch sticking forward from the base of one of the antlers (below)

Bison
Wisent

Bison bonasus

(Cattle Family)

Identification: HB 2.5-3 m, S 1.8-2 m, T 50-80 cm, W 500-850 kg; female noticeably smaller and lighter than male; coat colour reddish-brown in summer, blackish-brown in winter; hair on head and shoulders long and shaggy; tail tufted towards the end; both sexes have short curved horns; head carried rather low so that the shoulders are the highest part of the body.

Distribution: Once found throughout temperate parts of Europe, but became extinct in the wild in the early part of this century. Reintroduced and now well established in the Bialowieza Forest (Poland and Belorussia) and in the Caucasus.

Habitat: Deciduous and mixed woodlands with clearings and marshy areas.

Behaviour: Active by day and night; lives in small groups; adult males mostly solitary outside breeding season.

Food: Grass, herbs, twigs, fruit, bark.

Breeding: Mating season Aug-Sep; gestation 9 months; a single calf (rarely twins) born every 2 years; calf can walk after a few hours, is suckled for 6 months and independent at a year old.

This mighty animal once roamed the forests over the whole of Europe but retreated to the wildest forested areas as the land increasingly came under cultivation. At the beginning of this century the only remaining sites were in the marshy forest of Bialowieza in Poland and also in the Caucasus. In the latter site there was a mountain form of Bison living just below the tree-line but this became extinct in 1927.

The Polish Bison were decimated during World War I and they also became extinct, at least in the wild, in 1921. Luckily there were a few in zoos or private parks. From these captive animals the species was successfully reintroduced to Bialowieza, where it is now re-established with a population of some 2000 animals.

In summer, Bison usually graze during the morning and evening, retreating to the cool of the forest during the day to rest. In hot weather they drink at least twice a day and they also like to roll in dry soil or sand, but they do not wallow. During the spring moult they shed their hair in large clumps.

The Bison is very closely related to the **American Buffalo** (*Bison bison*) which was similarly almost exterminated in the wild before being rescued.

Lowland Bison (left) and Caucasus Bison (right)

Bison cow with calf (above)
Adult bull (below)

Aurochs

Bos primigenius

(Cattle Family)

Identification: HB 3-3.2 m, S 1.7-1.8 m, T 70-80 cm, W 600-1000 kg; female about a quarter smaller and lighter than the male; coat blackish-brown with pale stripe along back in male, reddish-brown in cow and calves; mouth has pale patch surrounding it; horns pale with sharp points.

Distribution: Once found from western Europe to eastern Asia and also North Africa; became extinct in Europe in seventeenth century.

Habitat: Open valley woodland with grassland and scrub, river deltas.

Behaviour: Lived in small herds under the leadership of a single animal; outside breeding season bulls roamed individually or in small groups.

Food: Grass, herbs, leaves, fruit.

Breeding: Mating season Sep-Oct; gestation about 9 months; usually a single calf; development as for domestic cattle, the cows reaching maturity at about 18 months.

Note: This now extinct species is the ancestor of domestic cattle.

Today the mighty Aurochs is represented only by references in literature, old paintings and bones. The last European Aurochs was killed in 1627 and by the eighteenth century the species was also extinct in Asia. In the 1920s, attempts were made in the zoos of Munich and Berlin to recreate the Aurochs by breeding programmes. By using primitive races such as Scottish highland cattle, Spanish fighting bulls and Hungarian steppe cattle, they were able to breed a race similar to the extinct Aurochs, although these animals didn't reach quite the size of the original.

We know surprisingly little about the biology of the Aurochs. Small family groups were probably led by an adult cow. In hard winters the herds probably migrated to find fresh supplies of food. Adult Aurochsen were apparently very fierce and agile, since hunting them was always considered dangerous. As early as 8000 years ago wild cattle were being domesticated and the Aurochs is the ancestor of all European races of cattle.

Size comparison between Aurochs (above, after a historical drawing) and a domestic breed (Simmental) (below). Selective 'back-breeding' in zoological gardens gives us an impression of what the long-extinct Aurochs might have looked like

Cow with calves (above)
Bulls (below)
These animals have been bred to resemble the Aurochs

Chamois

Rupicapra rupicapra

(Cattle Family)

Identification: HB 1.1-1.3 m, T 3-8 cm, W 30-50 kg; female somewhat smaller and lighter than male; resembles goat in overall body shape; in summer coat pale reddish-brown with black stripe along back and black legs, winter coat blackish-brown; black and white face markings; male and female both have a pair of fairly short horns that curve backwards.

Distribution: High mountains of central, eastern and southern Europe, also Turkey; has been introduced to a number of other mountain ranges in Europe.

Habitat: In summer occupies mountain grassland and scree slopes above the tree-line, spends the winter in mountain woodland.

Behaviour: Diurnal; climbs and jumps exceptionally well; lives in groups; older males solitary outside the breeding season; marks territory with scent glands on head.

Food: Grasses, herbs, leaves, shoots of coniferous trees; also bark, lichen and mosses.

Breeding: Mating season Oct-Dec; gestation about 6 months; single calf (rarely twins) which can follow its mother soon after it is born, suckled for 6 months and remains with the mother for about a year.

As in many members of this family both sexes of Chamois have horns. The horns of the male are somewhat thicker and more curved than those of the female. Horns are quite different in structure from antlers; they develop from bony protuberances under the skin of the forehead at an early age. The skin above the growing horns itself becomes horny, forming a thick hollow sheath. Unlike the antlers of deer, horns are not shed and regrown each year, but continue to grow from the base throughout the animal's life.

Male Chamois live solitary lives for most of the year but they come together during the rutting season in late autumn and strive to round up females. The opposing males stand side on to each other in a stiff-legged posture and try and intimidate one another. To exaggerate their size they also raise the hair along their backs. Short periods of interlocking horns are interspersed with chases over the rocks. Occasionally they lose their footing and slip, but serious wounds from the horns are rare. (Continued p. 216).

The typical black and white face pattern (left) fades with age (right)

Chamois can descend the steepest of gradients with great agility (above)
Female with young (below)

Chamois (continued)

Rutting male Chamois make loud grunting noises and they often mark their territories by rubbing their scent glands (behind the horns) on twigs or branches. This musk-like secretion probably also serves to excite the females. Rutting Chamois bucks use up a lot of energy and eat little, so at the end of the breeding season they are often tired and thin.

Before the hard winter weather sets in, Chamois usually descend to lower levels into the montane forest where they feed on shoots, bark and lichen. They often dig down under the snow to get at the grass. In summer they are much more selective in the food they take, and this includes young shoots and buds. Although they avoid certain plants such as ferns, mints or nettles, they can consume many species which are poisonous to people, for example yew, foxglove and deadly nightshade.

Chamois tend to rest in the middle of the day, settling down at a good vantage point to chew the cud. When they detect danger they make shrill alarm whistles by expelling air through their nostrils. Such alarm calls are also understood by Marmots which retreat rapidly into their burrows when they hear them. Female Chamois and their calves communicate with a goat-like bleating.

Chamois hooves are particularly well-adapted for their rocky environment. Their pads are comparatively soft and adhere to the stones to some extent, although each hoof also has a hard edge. When clambering downwards they also use the 'dew claws' behind the hooves and hardly ever lose their grip.

The chamois in northwest Spain, the Pyrenees and the Italian Apennines is rather distinctive – smaller, with more white on the throat and rump – and has been considered a separate species, the **Apennine Chamois** (*Rupicapra pyrenaica*).

Chamois marking territory using
scent glands on the head

Younger chamois in submissive gesture to dominant buck (above)
Chamois often have to dig beneath the snow to reach the grass (below)

Alpine Ibex

Capra ibex

(Cattle Family)

Identification: HB 1.1-1.45 m, T 12-15 cm, W 50-110 kg; male larger and heavier than female; grey-brown in summer, females and young greyer; in winter grey; horns large and curved backwards, in male up to 85 cm; rather flat front surface and with transverse ridges; female has smaller horns (to about 40 cm); male has short beard.

Distribution: The Alps.

Habitat: High mountains, rocks and alpine grassland above the tree-line.

Behaviour: Diurnal; males and females in separate herds outside the breeding season (herds consist of about 30 animals); climbs and jumps well.

Food: Grass, herbs, shrubs.

Breeding: Mating season Dec-Jan; gestation 5.5 months; normally has a single kid (rarely twins) which can follow its mother over the rocks within a few days and is suckled until the autumn; young females usually stay in a group with their mother and young males form single-sex groups when they are about 2-3 years old.

Since the Middle Ages Ibexes have featured in folk medicine and almost all parts of its body were thought to be a cure for a wide variety of illnesses. The species was so badly persecuted that by the beginning of the nineteenth century they were reduced to a single herd, in the Gran Paradiso National Park in Italy. All living Alpine Ibex are descendants of this one herd of about 50 animals. Since then they have been reintroduced to many alpine sites (there are more than 40 colonies in Switzerland alone) and wild Ibex probably now number more than 10,000.

The horns of an adult Ibex buck are truly impressive. Close inspection reveals dark horizontal bands which are annual rings indicating the age of the animal (these are not to be confused with the horizontal ridges). The dark bands result from a period of slow growth during the rut. During the winter rut the bucks chase each other. They measure their relative strengths in ritualised fights, in which they rear up on their hind legs and crash the bases of their horns together. Although such fights may seem fierce, they rarely result in serious injury. It is only during the breeding season that the adult males join the females. (Continued p. 220).

Adult Ibex can be aged by the narrow rings in their horns. This one is about nine years old.

Small group of young Ibex

Alpine Ibex (continued)

Alpine Ibex normally live high above the tree-line, and as the summer progresses they go to higher and higher levels to avoid the heat, descending again to lower slopes in winter. These winter quarters are usually steep, south-facing slopes that do not a deep covering of snow. Ibex are much heavier than Chamois (p. 214) and tend therefore to avoid snowdrifts, in which they are likely to sink. Like Chamois, Ibex have two claws on each foot in addition to the cloven hooves. These do not normally come into contact with the ground but are used when clambering down steep slopes.

The closely-related **Spanish Ibex** (*Capra pyrenaica*) is found in the Pyrenees, Sierra Nevada and central Spanish mountain ranges. It differs from the Alpine Ibex mainly in the shape of its horns, which are twisted in a slight spiral outwards and lack the transverse ridges. The male's horns can be up to a metre long but those of the female only measure about 20 cm. The Spanish Ibex is an endangered species because of its fragmented range; there are only small, isolated populations.

Fighting Ibex

Each tries with all its strength
to make the other give way

Alpine Ibex buck (above)
Spanish Ibex bucks (below)

Wild Goat

Capra aegagrus

(Cattle Family)

Identification: HB 1.2-1.5 m, T 15-20 cm, W 25-40 kg; horns sickle-shaped, curving backwards and sharply keeled along leading edge; male horns up to 1.3 m long, females about 20-30 cm; fur red-brown or grey-brown with a darker stripe along the back and across the shoulders; legs striped black and white; male has long beard.

Distribution: Greek islands including Crete, Turkey.

Habitat: Mountains between about 1200 m and 2200 m (more rarely up to 4000 m); steep, rocky slopes and scrub.

Behaviour: Active by day and at dawn and dusk; females and males form separate small herds outside the rutting season; older bucks solitary; climbs well.

Food: Grasses, herbs, tree foliage.

Breeding: Mating season Oct-Nov; gestation about 22 weeks; litter size 1-2; kids follow their mother after 4-5 days and are suckled for up to 6 months.

Note: This is the ancestor of domestic goats.

The Wild Goat has been exterminated over much of its former range in southwest Asia. It is also unclear whether the remnant populations are truly wild or whether they were deliberately introduced and have become feral. Wild populations also interbreed easily with domesticated goats. There is much dispute as to whether there are any pure Wild Goats left on Crete which at one time appeared to host the largest population of nearly purebred Wild Goats.

Human superstition is probably responsible for its drastic reduction in range. This species has long been hunted for the fur balls which are found in its stomach. These balls, which are also found in other goat species and in Chamois (p. 214), were thought to have marvellous medicinal properties. They consist largely of ingested hair which remains undigested in the stomach. They remain in the stomach and eventually mix with plant material to form stony, shiny balls. These hair balls were traditionally thought to cure sterility and to be effective against poisoning.

Fur balls from a goat's stomach

The Wild Goat shows typical adaptations to its rocky habitat. Males and females usually travel about in single-sex herds and individuals maintain a specific distance between each other when they rest in the middle of the day.

Goats probably began to be domesticated about 7000 BC and are therefore amongst the earliest of all domesticated animals. Small groups of feral goats (those derived from domestic stock but now living wild) occur in may places, for example north Wales, the Cheviot Hills and on many small islands.

Pair of Wild Goats

Mouflon

Ovis orientalis

(Cattle Family)

Identification: HB 1.1-1.3 m, T 6-10 cm, W 25-50 kg; male has tightly curled horns; females either lack horns altogether or have shorter horns; coat short and shiny; male reddish-brown with white saddle marking; female greyer; winter coat darker in both sexes.

Distribution: Originally from southwest Asia, introduced in prehistoric times to Corsica and Sardinia, now spread by introduction from there to other mountainous regions in Europe.

Habitat: Dry, stony mountain areas; also introduced to mixed woodland and parks.

Behaviour: Active by day and at dawn and dusk; lives in herds led by adult female; males stay in small groups outside the breeding season.

Food: Grass, herbs, fresh foliage, fruit, bark.

Breeding: Mating season Nov-Dec; gestation 5 months; usually a single lamb (occasionally twins) which follows its mother within a few hours and is suckled for 4-5 months.

Note: This is the smallest wild sheep species and the ancestor of domestic sheep. Those from Corsica and Sardinia were probably early domesticated stock that still resemble wild Mouflon more than modern domestic sheep.

Mouflon have been crossed with domesticated sheep in order to increase the hybrids' size for hunting. Mouflon lambs are born with a woolly coat that changes to the short adult coat at the first moult. Wild Sheep which retain their woolly coat for longer are probably of mixed parentage. Mouflon moult twice a year. Their winter coat is very thick and greasy. In their original rocky habitat, Mouflon travel over very hard, stony ground which wears down their hooves very quickly. In areas where they have been introduced to softer ground their hooves often grow very long and give them walking problems.

Mouflon lambs start to grow their horns soon after they are born and their horns grow very rapidly for the first four years, more slowly thereafter. By the time they are nine years old the characteristically curled horns are fully developed. The largest horns can weigh up to 5 kg. During the rut, the sounds of Mouflon horns clashing together can be heard over long distances. The rams are protected from concussion by their thickened skull bones.

Mouflon ram: the arrows point to the annual growth rings which can be used to calculate the age of the animal

Mouflon ram (above)
Ewe with two lambs (below)

Musk Ox

Ovibos moschatus

(Cattle Family)

Identification: HB 1.6-2.5 m, S 1-1.3 m, T 8-15 cm, W 200-400 kg; male larger and heavier than female; resembles small cow; fur very long, dark brown with lighter saddle markings; legs yellowish-grey; horns drooping to either side of head and curving upwards, growing from broad bases on crown; hooves wide and spreading; calves bleat, males bellow in mating season.

Distribution: Arctic North America and Greenland; introduced to Norway and Spitzbergen.

Habitat: Arctic tundra.

Behaviour: Diurnal, also nocturnal in winter (during the long Arctic night); herds of between 10 and 20 animals gather, in winter these grow up to 100; mainly sedentary.

Food: Grasses, herbs, mosses, dwarf shrubs, lichen.

Breeding: Mating season Aug-Oct; gestation about 8.5 months; usually a single calf (rarely twins) every 2-3 years; calf follows the mother after a few hours, is suckled for about a year, independent at 2 and mature at 4 years old.

Musk Oxen combine features of sheep, goats and cattle, as reflected in their generic name *Ovibos* (meaning sheep ox). The species was named Musk Ox soon after its discovery in 1720 because of the strong musky odour which the bulls emit during the breeding season. (The musk used in the perfume industry, however, comes from the Chinese Musk Deer.)

During the last Ice Age, Musk Oxen were also found in central Europe. As the ice retreated so the species moved back with it to the north of Europe – they cannot tolerate warm conditions. Their thick coats protect them from the fiercest Arctic storms and they have the longest hair of any wild animal, reaching 90 cm on the neck and flanks. In bad weather the members of the herd huddle close together, protecting calves in the middle of the group. They also adopt a similar formation when danger threatens, for example in the case of attack by a Wolf pack or by Polar Bears. The attacking animal is then confronted by a circle of adult Musk Oxen, each equipped with sharp horns.

Nevertheless Musk Oxen are easy prey to human hunters and were exterminated in the Eurasian Arctic. The species has been brought back from the verge of extinction through reintroductions and careful protection and numbers now stand at around 24,000 worldwide.

Musk Oxen in defensive formation

Musk Ox bull (right) with cows (above)
Moulting out of winter coat (below)

Marine mammals – relative sizes

Humpback Whale

Minke Whale

Northern Right Whale

Blue Whale

Sperm Whale

Common Dolphin

Common Seal

Narwhal

Monk Seal

Killer Whale

Walrus

Beluga

Common Dolphin

Delphinus delphis

(Dolphin Family)

Identification: L 1.8-2.2 m, W 75-115 kg; slender and streamlined; dark grey above, whitish below, with pale grey, yellowish or white patch along side; dorsal fin triangular and somewhat curved backwards.

Distribution: Worldwide, in tropical and warm seas, the Mediterranean Sea and Black Sea; follows the north Atlantic currents as far as British Isles and Scandinavia and occasionally into the North Sea. Mainly towards the south in British waters.

Habitat: Prefers coastal waters.

Behaviour: Sociable, living in large groups; often jumps out of the water; communicates using a series of whistles; navigates and detects prey by echolocation.

Food: Fish (mainly surface water species), squid.

Breeding: Mating season June-Sep; gestation 10-11 months; a single calf (rarely twins) which is suckled for more than a year.

Dolphins, porpoises and whales together comprise the cetaceans. All of the larger species are known as whales although some, like the Killer Whale, are members of the dolphin family, all of which have numerous pointed teeth in both the upper and lower jaws. The Common Dolphin is the commonest and best-known member of the toothed whales. Whales and dolphins have evolved from land animals which returned to the water more than 60 million years ago. They gradually became better and better adapted to their aquatic habitat and now have many superficially fish-like characteristics. Their hind limbs have disappeared altogether and their front limbs are modified as flippers. They swim by moving the large tail flukes powerfully up and down.

Dolphins usually swim in groups or schools of between 15 and 30 individuals, although they are sometimes found congregating in their hundreds. They are highly inquisitive and sociable animals and can often be seen following ships and making repeated jumps out of the water. Dolphins of many species are increasingly being drowned by accidental capture in fishing nets, as has happened with Common Dolphins in mackerel nets off Cornwall.

A dolphin's mouth is equipped with rows of pointed teeth, a distinctive feature of members of the toothed whale group. The teeth number 80-100 (in rare cases up to 250), a record for a mammal. Each tooth curves slightly backwards and this helps them catch and swallow their fast-moving fish prey (species such as herring, mackerel and mullet). Toothed whales use an echolocation system to help them navigate and also to track down their prey. This is in many ways comparable to the sonar employed by bats. They produce a rapid series of clicks, mainly in the ultrasonic range and therefore inaudible to people. Analysis of the echoes allows them to build up a precise 'sound picture' of their environment. They also make a whole series of whistles and quacking sounds to communicate with one another.

Two other southern dolphins the **Striped Dolphin** (*Stenella caeruleoalba*) and the **Rough-toothed Dolphin** (*Steno bredanensis*) occur in southwest Europe.

Dolphins often jump out of the water together

Bottle-nosed Dolphin

Tursiops truncatus

(Dolphin Family)

Identification: L 1.75-3.7 m, W 150-200 kg; dark grey above, pale grey or white below; beak short (bottle-nose); lower jaw protruding; dorsal fin large and triangular, curved backwards.

Distribution: Worldwide, in warm and temperate seas; Mediterranean Sea, Black Sea, North Sea, more rarely into the Baltic Sea.

Habitat: Coastal waters.

Behaviour: Sociable, mainly in groups (schools) of 5-10, sometimes up to 100 animals; males and females live separately outside the breeding season; often jump out of the water, sometimes several individuals at once; most European populations appear to be sedentary, with only a little offshore–onshore migration; navigation and prey location by ultrasonic clicks (echolocation).

Food: Fish (mainly herring, mackerel and other schooling species), also squid.

Breeding: Mating season spring–summer; gestation about 11 months; a single well-developed young calf (rarely twins), suckled for 18 months to 2 years.

Note: This species is easily tamed and can be taught to perform tricks.

This is the dolphin most commonly seen in dolphinaria and which also stars in the television series Flipper. Like Common Dolphins, Bottle-nosed Dolphins often follow ships at sea.

All whales and dolphins give birth underwater and the calves are born tail first. The newborn calf must swim up to the surface for its first gulp of air and the mother often helps it in this process. Newborn Bottle-nosed Dolphins are almost half the

Bottle-nosed Dolphin giving birth

size of the adult and are already well-developed. This is important since they must be able to swim immediately and follow their mother. The calves also have a good layer of insulating fat to protect them from the cold. The mother dolphin suckles her calf for up to two years and protects it from danger.

Close observation of Bottle-nosed Dolphins has confirmed some surprising stories about dolphin behaviour. If an individual is ill or injured, other dolphins may actually go to its aid and try and save it from drowning. Two healthy dolphins support the injured animal, one on either side, and push it upwards until it reaches the surface where it can breathe air. There are also many stories of dolphins helping to save people from drowning in a similar fashion.

Living close to the shore, this species is very vulnerable to pollution and has declined considerably in British waters in the last few decades.

Two other species of dolphin are found in the seas between Britain, Iceland and Norway: the **White-beaked Dolphin** (*Lagenorhynchus albirostris*) and the **White-sided Dolphin** (*Lagenorhynchus acutus*).

Leaping above the water surface (above)
Breathing, note open blow-hole (below)

Killer Whale
Orca

(Dolphin Family)

Orcinus orca

Identification: L 4.5-9 m, W 2500-4500 kg; male larger and heavier than female; back and sides black, underside white with an oval white patch behind the eye; dorsal fin long and narrow, up to 1.8 m tall in male.

Distribution: Worldwide, from the Arctic to the Antarctic Oceans, regular along the Atlantic coasts of Europe, sometimes found in Baltic and Mediterranean seas.

Habitat: Prefers colder water.

Behaviour: Lives in family groups (schools) or 5-20 animals, often hunting in co-ordinated packs; swimming speed up to 55 km per hour; navigation by echolocation.

Food: Large fish, sea birds, seals, other species of whale.

Breeding: Mating season throughout the year; gestation 14-15 months; a single calf 2 m long at birth, which is suckled for a year.

The black and white Killer Whale is the largest of the dolphin family. The long, narrow dorsal fin is almost vertical in adult males, but curves gently backwards in females and young animals. The fin is not supported by a bone but is kept upright by blood pressure.

Dorsal fins of male (left), female (centre) and young (right)

The Killer Whale is the only species of whale or dolphin which regularly feeds on other sea mammals, including other whales. Killer Whales have a somewhat unjust reputation for ferocity, in fact they only kill for food. When hunting they often make co-ordinated attacks, particularly when the prey is a large whale. They can inflict such serious wounds on larger whales that the prey is gradually weakened by blood loss and eventually overpowered. Nevertheless Killer Whales sometimes allow dolphins and other whales to swim amongst them unharmed. Killer Whales will occasionally attack seals as they lie on pack ice. Two Killer Whales will heave up the ice from below, sliding the seal towards a third Killer Whale waiting in the water. They often use a similar tactic when hunting penguins.

Perhaps surprisingly, Killer Whales can be easily trained and display apparent affection in captivity although like all captive wild animals they are prone to disease and show unnatural behaviour.

The **Long-finned Pilot Whale** (*Globicephala melas*) is another very large member of the dolphin family – up to 6 m– found all along the Atlantic coasts of Europe.

Female (left) and male (right) in their natural habitat (above)
Killer Whale showing sharp curved teeth (below)

Risso's Dolphin

Grampus griseus

(Dolphin Family)

Identification: L 3.3-3.8 m, W 300-400 kg; a large dolphin with blunt head, no beak and pointed dorsal fin. Uniform grey above, usually with conspicuous lighter scars.

Distribution: Worldwide except in Polar seas; European coasts from the Baltic, round the British Isles to the Mediterranean. Most easily seen off western Scotland and western Ireland.

Habitat: Coastal waters, usually within 10 km of land.

Behaviour: Usually in small groups up to 10, rarely in groups of up to 50. Young sometimes leap clear of the water; sometimes lie vertically in water with head protruding.

Food: Mainly squid, cuttlefish and octopus; some fish.

Breeding: Little currently known; young born in spring and early summer.

Porpoise

Phocoena phocoena

(Porpoise Family)

Identification: L 1.4-1.7 m, W 55-65 kg; the smallest cetacean; no beak, dorsal fin low and blunt. Teeth flattened rather than conical like all the dolphins.

Distribution: Temperate north Pacific and north Atlantic oceans, and in all European seas including the Baltic, Mediterranean (now very rare) and Black.

Habitat: Coastal waters including estuaries and river-mouths.

Behaviour: Lives in small groups of up to 10; never jumps clear of the water and therefore difficult to detect unless the water is calm.

Food: Fish, especially herring, crustaceans and cuttlefish.

Breeding: Mating in summer, gestation 11 months, single young born between May and July. Weaned when about 8 months old.

Porpoises are very vulnerable to accidental capture and to pollution. They have declined drastically in recent decades, especially in the Baltic Sea, southern North Sea, Irish Sea and Mediterranean Sea.

Risso's Dolphin (above)
Porpoise (below)

Beluga
White Whale

Delphinapterus leucas

(White Whale Family)

Identification: L 3.7-5 m, W 800-1200 kg; male larger and heavier than female; entirely milky white, calves brown or grey; lacks dorsal fin.

Distribution: Northern Arctic Ocean, subarctic seas.

Habitat: Coastal waters, bays, fjords, occasionally in estuaries.

Behaviour: Sociable, lives in groups of 5-10 animals and will occasionally gather together in larger herds (up to 1000); swims slowly; uses echolocation.

Food: Fish (mainly schooling species such as cod, herring and salmon); squid; crustaceans.

Breeding: Mating season Apr-June; gestation about 12 months; a single calf (rarely twins) born every 3 years and suckled for 2 years.

Its pure white body makes this species quite unmistakable. Calves, however, are brown at first, changing through steely-grey, then becoming speckled, and eventually pure white.

The Beluga's flexible neck (unique amongst whales) allows it to turn its head from side to side. Belugas are found mostly in shallow water and are regularly seen off the north coast of Norway. They occasionally come south as far as the Baltic and North seas. Belugas are very vocal whales and their calls can often be heard above the water.

Narwhal

Monodon monoceros

(White Whale Family)

Identification: L 4-5 m, W 600-1000 kg; upper side with irregular grey speckles, below uniform dark or pale grey; dorsal fin lacking; male has long straight tusk, 1.8-3m in length.

Distribution: North Arctic, only rarely south to Iceland and northern Norway; very exceptionally to British waters.

Habitat: Shallow coastal waters, fjords.

Behaviour: Lives sociably in herds which are subdivided into groups of females with their young and groups of young males; swims rapidly and can dive for up to 30 minutes.

Food: Fish (e.g. polar cod), squid, cuttlefish, crustaceans.

Breeding: Mating season Apr-June; gestation about a year; a single calf (rarely twins) born every 3 years and suckled for 2 years.

The straight, spirally grooved tusk of the male Narwhal is highly characteristic. This object was long revered as the legendary Unicorn's horn and was said to possess magical powers. Narwhals have only two teeth which grow horizontally in the upper jaw. In the female these usually remain in the gums but in the male the left tooth grows out to form the tusk, which may be up to 3 m long. In rare cases (about 1 in 100 males) both teeth develop into tusks. For a long time it was thought that Narwhals used their tusks to pierce holes in the ice or to dig in the seabed for food. However, this did not explain why it was only the male which developed the tusk. Observations from the air have now confirmed that rival males use their tusks in hierarchical disputes and when contesting rights to the females.

Belugas often call above water (above)
School of Narwhals in formation (below)

Sperm Whale

Physeter macrocephalus

(Sperm Whale Family)

Identification: L male 15-25 m, female 9-12 m, W male 45-70 t, female 15-20 t; dark grey with whitish underside; head very large, blunt-ended and barrel-shaped; dorsal fin very small and extended as ridge towards tail; flippers small; tail flukes up to 4 m across.

Distribution: Worldwide in tropical and subtropical seas; adult males migrating into temperate waters in summer.

Habitat: High seas.

Behaviour: Lives in groups (schools) usually of between 10 and 30 individuals; males often solitary outside breeding season; dives deep and for long periods; uses echolocation to detect prey and for navigation.

Food: Almost exclusively squid, but fish are taken in some areas.

Breeding: Mating season in northern hemisphere usually June; gestation 14-15 months; single calf (rarely twins) born every 3 years and suckled for about 2 years.

The mighty Sperm Whale is the largest of all the toothed whales. In the eighteenth and nineteenth centuries this species was the whale most favoured by the whaling industry. A white variety of the species is also famous as the title role in Herman Melville's novel Moby Dick.

The Sperm Whale's enormous head, which accounts for a third of the body length, contains an oily substance which solidifies to a wax when exposed to the air. This substance, known as spermaceti wax, was once valued as a raw material for making candles, cosmetics and medicinal creams. The function of the spermaceti organ is to act as a kind of acoustic lens for focusing sound through the layers of wax, but also probably as a buoyancy regulation mechanism. By using this organ Sperm Whales can navigate successfully and hunt prey in the darkness of the ocean depths.

The favourite food of Sperm Whales are squid which they catch at depths of up to 1200 m. In deep dives Sperm Whales can remain submerged for more than an hour. Sperm Whales have a distinctive signature when blowing (exhaling air) at the sea surface. The blow of a Sperm Whale projects forwards at an angle of 45°.

Another valuable product of the Sperm Whale is ambergris, which is an intestinal secretion. This is used in the perfume industry to bind scents. Ambergris is occasionally found in large lumps in Sperm Whale intestines. Sperm Whales have been protected from commercial exploitation since 1985.

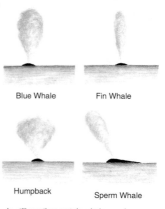

Blue Whale

Fin Whale

Humpback

Sperm Whale

In still weather certain whale species can be identified by the shape of their blow

Sperm Whales have rows of conical teeth, but only in the lower jaw (above)
Female Sperm Whale with calf (below)

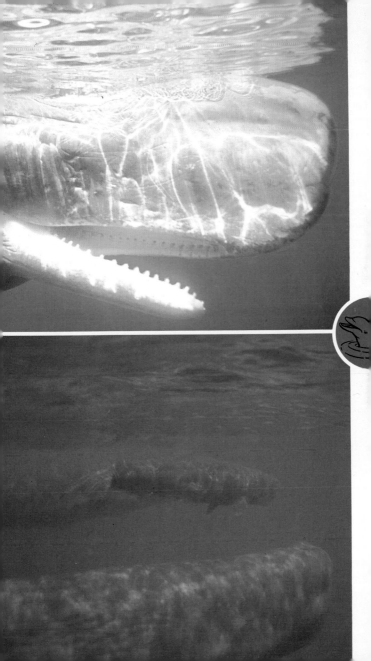

Blue Whale

Balaenoptera musculus

(Rorqual Family)

Identification: L 20-30 m, W 120-150 t; female somewhat larger than male; colouring blue-grey with irregular white speckles; flippers narrow and pointed; dorsal fin very small; 70-120 longitudinal furrows along throat; baleen plates (see below) blue-black; blow vertical, 6-9 m high (see p. 240).

Distribution: Worldwide in all major oceans. Now very rare in the east Atlantic.

Habitat: In summer prefers cold water towards the Arctic and Antarctic, moving towards more temperate and warmer waters in the winter.

Behaviour: Either solitary, in pairs or small family groups; dives for up to 50 minutes, breathing 10-15 times at the surface between dives.

Food: Almost exclusively small planktonic crustaceans (especially krill).

Breeding: Mating season in the northern hemisphere usually Dec-Jan; gestation about 11 months; a single calf (rarely twins) born every 2 years and suckled for 6-7 months.

Note: Largest living animal, and largest animal ever known to have lived on Earth, weighing as much as 30 elephants.

Even a Blue Whale calf is more than 7 m long at birth and weighs about two tonnes. The calf feeds at first on its mother's milk, putting on almost 100 kg a day. By the time it is weaned at six or seven months old it weighs almost 25 tonnes.

The Blue Whale belongs, along with the Fin Whale (p. 244) and Humpback Whale (P. 246), to the group known as the baleen whales. All Baleen whales depend upon tiny prey which live in the oceanic plankton. Of primary importance are the krill, matchstick-sized swimming shrimps which appear in huge numbers in the cold polar seas. A Blue Whale will eat about four tonnes a day, comprising around four million shrimps.

Instead of teeth, the baleen whales are equipped with so-called baleen plates. These act as filters to sift the edible plankton from the seawater. The plates hang like curtains from the upper jaw and each has a brush-like fringe. When feeding, a whale gulps a large quantity of sea water containing plankton. It then closes its mouth and by pressing its thick tongue upwards, forces the water to either side through the baleen plates. Krill and other plankton are filtered out on to the bristles and are then swallowed.

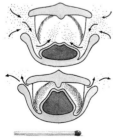

Cross-section of mouth of baleen whale. The whale opens its mouth, taking in sea water and plankton (above). It then shuts its mouth, pressing upwards with its thick tongue (dark grey), sieving the water through the baleen plates (below). Krill and other plankton are caught on the plates (white) and are then swallowed

Krill, the main food of the Blue Whale

Blue Whale breaking the surface (above) and diving (below)

Fin Whale

Balaenoptera physalus

(Rorqual Family)

Identification: L 18-25 m, W 40-52 t; female somewhat larger than male; grey above, white below; flippers narrow and pointed; dorsal fin small and triangular; head long and wedge-shaped; 80-100 throat grooves; front baleen plates (see p. 242) yellowish-white, inner plates blue-grey.

Distribution: Found worldwide in all large oceans; in the Atlantic commonest around Iceland, Faroes and Norway; also west of the British Isles and in west of the Mediterranean Sea.

Habitat: Mainly the high seas.

Behaviour: Irregular north–south migrations; usually moves around in small groups of 2-6, but in rich feeding sites herds of up to 100 may congregate; feeds mostly at the surface or at a shallow depth; the fastest of the large whales, reaching speeds of up to 33 km per hour.

Food: Small crustaceans, also herring and other small shoaling fish.

Breeding: Mating season Oct-Mar; gestation 11-12 months; a single calf (rarely twins) every 2 years, suckled for about 6 months.

The head of the Fin Whale is irregularly marked. On the right hand side, the lower jaw is white and the mouth grey but on the other side the colouring is the other way round. The whale turns on its side when feeding, keeping the pale side towards its prey. The throat grooves reach right back to the belly region and these have an important function during the feeding process in all baleen whales. The whale takes in a large volume of water into its mouth and during this process the throat grooves allow for a rapid expansion of volume.

The furrowed throat allows a rapid inflation of the mouth cavity

Fin Whales are no longer being hunted.

The Right Whale family also belongs to the baleen whale group. The **Bowhead Whale** (*Balaena mysticetus*) and the **Northern Right Whale** (*Eubalaena glacialis*) both occur off the coasts of northern Europe. Both these species are threatened through over hunting.

The very similar **Sei Whale** (*Balaenoptera borealis*) also occurs in the eastern north Atlantic, and the much smaller **Minke Whale** (*Balaenoptera acutorostrata*) is the most abundant baleen whale in European waters. The Minke Whale is the only whale now being hunted commercially in Europe, by the Norwegians.

Fin Whales have a small triangular dorsal fin (above)
Fin Whales sometimes throw themselves right
out of the water, known as breaching (below)

Humpback Whale

Megaptera novaeangliae

(Rorqual Family)

Identification: L 12-18 m, W 25-33 t; female somewhat larger than male; black above, speckled white below; flippers narrow and very long; dorsal fin small; head and edges of flippers covered with small lumps; trailing edge of tail uneven; 15-40 throat grooves; baleen plates black; blow pattern very wide and only 2-3 m high (see p. 240).

Distribution: Worldwide in the larger oceans; very rare in the east Atlantic.

Habitat: Coastal waters with plenty of fish and plankton.

Behaviour: Lives in family groups of 3-20 individuals; swims very slowly; occasionally leaps right out of the water (breaching); regular yearly migrations between polar feeding waters and subtropical or tropical wintering waters; complex underwater songs.

Food: Planktonic crustaceans, small schooling fish, squid.

Breeding: Mating season (northern hemisphere) April; gestation 10 months; a single calf (rarely twins) every 2 years; suckled for at least 6 months.

The Humpback Whale is one of the slowest whales, with a normal swimming speed of only 3-8 km per hour. This slow speed, combined with its coastal distribution make it an easy target for whalers. Although numbers have now recovered somewhat since commercial whaling ceased in 1965, the species is still classified as vulnerable.

Humpback Whales follow traditional routes on their north–south migrations. North Atlantic populations travel between Greenland and Iceland in the spring into the Arctic Ocean around Spitzbergen, returning in the autumn southwards along the Norwegian coast, past the British Isles, towards the west coast of Spain.

The large baleen whales probably do not use an echolocation system like the toothed whales (see p. 230). They are nevertheless very vocal, particularly on the breeding grounds, where Humpback Whales are well known for their varied underwater singing. Indeed the songs of Humpback Whales have even become famous as commercial recordings. Individual whales seem not to have a particular song which they learn and perfect like birds, but rather they 'compose' new songs each season. The complex songs probably enable subtle communication between individuals. Humpback Whales usually do not all sing at once, but alternate as in a conversation.

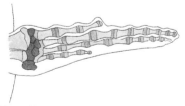

Flipper of Humpback Whale showing skeleton

Humpback Whale with open mouth showing baleen plates (above)
Despite their weight of up to 30 tonnes, Humpback
Whales can leap clear of the water (below)

Mammal Biology

What are mammals?

Mammals are vertebrates, a group that also includes fish, amphibians, reptiles and birds. There are some 4500 species of mammal, including our own species, about 200 of which are found in Europe. They show extreme variation in shape and size, but share certain characteristics.

- They are warm-blooded, in that their body temperature is not usually dependent upon that of their environment.

- They have body hair, modified as spines in some species (hedgehog, porcupine, spiny mouse), or reduced (whales and dolphins and man). Hair not only gives good insulation from the cold, but also allows camouflage or signal patterns. Special elongated sensitive hairs serve as tactile bristles.

- Young mammals grow inside the body of their mother and are born relatively well developed, although this varies from group to group. In Australia and New Guinea there are egg-laying mammals (Duck-billed Platypus and echidnas), but these are exceptions. In some groups (e.g. insectivores, rodents and carnivores) the young are born blind and often naked – fully dependent at first upon maternal care. Other species are born so advanced that they can stand and run or swim soon after birth (e.g. ungulates, hares and whales).

- Female mammals suckle their young on milk – a highly nutritious secretion produced in the mammary glands of the chest or belly.

Why are wild mammals difficult to spot?

We see birds everyday, often in large numbers, and the same is true of insects. So why not mammals? Where are the mice, badgers, deer, otters, hares, dormice and bison? In some cases the answer is that they have become so very rare, but the main reason is that mammals have become experts in concealment. Unlike birds, they cannot easily escape danger, so the strategy most mammals adopt is one of danger avoidance.

Many mammals rely on the protection of the dark, being nocturnal or crepuscular (active only at dawn or dusk) and resting or sleeping in the relative safety of their dens, nests or other cover by day. Others are protected through their camouflaged fur patterns which resemble their environment – for example the dappled coat of a Roe Deer fawn or the earth-coloured fur of the Brown Hare. Some mammals (e.g. Stoat, Mountain Hare, Arctic Fox) that live in areas with regularly snowy winters change into a white coat each autumn and are then harder to spot against the snow and ice. Even the best camouflage could be betrayed by the slightest movement and so many mammals combine camouflage with standing absolutely still, or lying motionless on the ground, jumping only at the last moment into a sudden dash for safety. Other species flee at the first sign of danger.

Watching mammals requires a special technique, as well as considerable patience. Rather than tramping through a habitat, one needs to sit perfectly still in one place and watch carefully. It is often amazing how many species it is possible to observe in this way.

Mammals as specialists

During the course of their evolution mammals have adapted to a huge range of habitats and conditions, becoming specialised in different ways, and these specialisations are reflected in a wide range of body shapes.

Bodies, limbs and feet

The basic mammalian limb is used in many different ways: bats have evolved wings; seals, dolphins and whales have flippers; moles have short, powerful shovels. Swimming species, such as beaver and otter, have webbed toes. Yet another development is the hard horny feet and long, narrow legs of the hoofed mammals – an adaptation to rapid, sustained running.

Mammals may be grouped according to which parts of their feet they use in locomotion. Those that walk on the soles of their feet are known as **plantigrade** (e.g. bears, some members of the weasel family, insectivores and apes); the **digitigrade** species (e.g. cats and dogs) walk on their toes. The hoofed mammals are **unguligrade**: they have taken the digitigrade plan one step further and use just the enlarged tips of their toes for walking and running.

In the **odd-toed hoofed mammals** (e.g. horses) it is the much enlarged third toe which takes the weight of the body, the other toes being vestigial.

In the **even-toed hoofed mammals** (e.g. pigs, deer and bovids) on the other hand, their cloven hooves consist of modified third and fourth digits. In deer and in most bovids the second and fifth toes are visible as reduced side toes higher up the foot, known as dew-claws, and in pigs these sometimes help support the foot as well, although they are usually carried clear of the ground.

Aquatic mammals such as whales, dolphins, seals and otters have streamlined bodies which help them swim efficiently, whilst those species that live largely in the soil (e.g. voles and moles) have reduced their external appendages to a minimum and developed cylindrical body shapes.

Eyes

Eye position is similarly related to biology. Many mammals that swim (e.g. beaver, Coypu, Otter) have eyes placed high on their heads so they can scan their environs even when swimming, without having to raise their heads out of the water.

Carnivores have both eyes facing forwards so that they can focus accurately on their prey using binocular vision. Herbivores by contrast do not need to see their food particularly well; for them it is much more important to be able to spot any signs of imminent danger, even whilst feeding. Thus in hoofed mammals, rodents and lagomorphs the eyes are set at the sides of the head, and may even bulge noticeably (e.g. mice).

The eyes of hares are at the top of the head, giving all-round vision, including the sky above so they can spot birds of prey without having to move.

Teeth and jaws

Number and shape of the teeth are also highly typical for each species of mammal, and important as taxonomic characters for experts. But even an amateur can tell the diet of an animal from its skull, and often identify the group to which it belongs.

Insectivores have rows of more or less similar sharply pointed teeth with which they grip and crack open the shiny exoskeletons of their insect prey. Bats, which also eat mainly insects, have a similar arrangement of teeth.

Rodents and lagomorphs have large chisel-like incisors, and a large gap separating these from the broad-crowned molars at the back. The incisors are used for gnawing plant material which is then ground down by the molars.

The jaws of a carnivore are quite different. The dagger-like canines help them catch and kill even quite large prey, and their molars are not flat, but sharply ridged to help them rip up flesh.

Ruminants (deer and bovids) lack incisors in the upper jaw, and there is a long gap between their front and back teeth. The upper palate is hard and ridged to help break up the food, and their molars are flat and broad, typical for a plant-eater.

Omnivores (e.g. Badger, bears) combine features of both carnivore and herbivore dentition, with sharp canines and broad molars.

The baleen whales have a highly unusual method of feeding, reflected in their mouth parts, which are specialised for filtering (see p. 242).

How mammals deal with the cold

Mammals that live in cold or temperate regions need to expend much energy in maintaining a constant high body temperature. Thick fur and underfur is one way of protection from cold air, and polar mammals such as Musk Ox and Polar Bear exemplify this strategy. Many mammals from more temperate latitudes develop a thicker winter coat.

Aquatic mammals are at particular risk from the cold. They tend to have thick layers of fat below the skin (known as blubber in seals and whales), or they smear their fur with oily secretions to prevent waterlogging (e.g. Otter, beavers). In the latter case a layer of insulating air is trapped in the underfur.

Many mammals retreat to their burrows in cold weather. They become much less active, saving energy at a time of food shortage, and use up food reserves laid down in their bodies, or stored in their nests.

Mammals (e.g. bears, Badger, squirrels) that spend much of the winter or cold period asleep in a safe site are said to show a period of dormancy. Some enter a state of true hibernation (e.g. dormice, marmots, sousliks, bats) in which pulse rate, breathing rate and body temperature are reduced to a minimum (see p. 60). In order to ensure that their young are born at the best time of year (normally in spring or early summer) while minimising breeding-related activity in winter, some species have developed 'delayed implantation' of the embryo. In the Stoat, for example, mating takes place in the summer, but the fertilised egg only undergoes a very small amount of development and then remains dormant throughout the winter. It then becomes implanted in the wall of the uterus in early spring and only then resumes development until the foetus is ready for birth. A similar strategy is found in several other carnivores and in Roe Deer. In bats there is a variation of the strategy – following mating in the autumn or winter, the sperm is stored in the uterus until it is required when ovulation takes place in the spring.

Mammals and people

People have long influenced their environment and this obviously includes and affects the wild mammals with which they share it. Some species were domesticated, others introduced to Europe from abroad, either for sport, food or for fur-farming (e.g. Muskrat, Fallow Deer, Axis Deer, Sika Deer). Fur-bearing species are (or have been) kept in fur farms, from which individuals escaped from time to time. Some of these managed to establish themselves and now form part of the wild mammal fauna of Europe (e.g. Coypu, American Mink, Raccoon, Raccoon Dog).

Certain species are regarded as undesirable or pests and have been much persecuted (e.g. rats, mice, carnivores). The Wild Horse and Aurochs were both driven to extinction by over-hunting, and the Bison only rescued from the same fate through re-introduction from captive stock. Whaling nations drove the larger whale species towards extinction although some species have now begun to recover since hunting of the larger species was discontinued.

Many wild mammals have lost out as Europe has become more industrialised and urbanised, and we continue to destroy more and more habitats, restricting wild mammal species to ever decreasing suitable sites. The shyer species, such as Otter, European Mink, Lynx and Wild Cat, are particularly sensitive to such pressures. On the other hand, cultivated and suburban areas have opened up new habitats for some mammals, such as fox, Beech Marten, Brown Hare, House Mouse, and some species of bat.

A more recent threat comes from pollution of the air and water with substances including chemical effluent and pesticides. There are now signs that people are deeply concerned about such pollution and in many areas steps are being taken to reduce it. There is also a growing awareness of the need to conserve threatened species, and this is supported by publicity from conservation bodies (see p. 252). Laws also prevent trade in certain species, and the following European species are fully protected: Brown Bear, Otter, Wild Cat, Lynx, Monk Seal, Przewalski's Horse, Mouflon, Apennine Chamois and most species of whale.

Some organisations involved in mammal conservation

WWF UK (World Wide Fund for Nature)
Panda House
Weyside Park, Godalming
Surrey GU7 1XR

Mammal Society
15 Cloisters Business Centre
8 Battersea Park Road
London SW8 4BG

Fauna and Flora Preservation Society
1 Kensington Gore
London SW7 2AR

There are also many local groups and organisations throughout Britain dedicated to protecting wild animals. Mammal conservation should not be the concern solely of these organisations however, and we can all play our part in helping to protect threatened species.

This guide should help to deepen our understanding of European mammals. Our understanding and care should help increase the chances of their survival in the future.

Photographs:
Front cover: Weber. Back cover: Reinhard (otter), Danegger (fox), Meyers (hare), Weber (badger). Insides: Angermayer:155 above, 157 below, 211 above, 213 above, 213 below, 217 below, 223; Arndt: 79 above, 79 below, 165 above, 179 below, 189 above, 211 below; Bender: 125 below, 153 above, 167 above, 189 below, 199 below, 205 below; Brandl: 201 below (FLPA); Bruemmer: 185 below, 239 below, 239; Clark: 41 below (FLPA); Cramm: 129 above; Danegger: 11 below, 59 above, 59 under, 61 above, 113 below, 115 below, 121 above right, 161 above, 161 below, 163 above, 191 above, 193 above, 193 middle, 193 below, 195 above, 199 above, 207 above, 215 above, 215 below, 217 above; Eichhorn/Zingel: 159, 201 above; Frey: 245 below; Gröndal: 209 below; Jacana/Gohier: 231, 235 above, 235 below, 243 above, 243 below, 247 above; C. König: 31 above right, 35 below, 39 above; R. König: 65 above, 65 below, 71 below, 197 below, 225 below; Labhardt: 173 above; Layer: 69 above, 73 above, 85 above, 101 below, 175 above; Lehmann: 177 below; Limbrunner: 21 above, 29, 31 above, 33 above right, 35 above, 35 middle, 39 below, 41 above left, 41 above right, 43, 45 above, 45 below, 47 above, 47 below 63 below, 91, 93 above, 101 above, 101 middle; Löhr: 67 above; Maier: 85 below, 133 above, 139 above, 147 above, 165 middle; Meyers: 2/3, 61 below, 113 above, 147 below, 163 below, 195 below; Ochotta: 109; Okapia: 57 above, 99, 111 above, 181 below (Davis); Pforr: 13 above, 15 above, 15 below, 17 below, 71 above, 73 below; Pitman/Earthviews: 237 above; Pott: 165 below, 207 below, 225 above; Quedens: 77 above; Reinhard: 9 below, 11 above, 19, 53, 55 above, 57 below, 81 above, 81 below, 87 above, 97, 121 above left, 121 below, 123 above, 127 above, 127 below, 131 above, 131 below, 137 below, 141 above, 141 below, 143, 155 below, 167 below, 169 below, 173 below, 197 above, 205 above; Reinhard/Angermayer: 90, 87 below left, 89, 103 below, 131 middle, 133 below, 139 below; Sauer: 111 below, 233 above; Siegel: 33 above left, 37; Silvestris: 17 above, 33 below, 23 above (Wilke), 23 below (Höfels), 49 above (Nill), 49 below, 67 below (Varin), 95 above, 95 below, 125 above, 149 above (Lane), 149 below (Lane), 151 above (Rohdich), 157 above (Höfels), 181 above (Coleman), 183 (Coleman), 191 below (Meyers), 233 below, 241 above, 241 below (Coleman), 245 above, 247 below (Lane); Skogstad: 187 above, 187 below; Schrempp: 151 below, 203 below; Schumann: 129 below; Schwammberger: 13 above, 75 above, 75 below, 107 above, 107 below; Steinberger: 277 above; Stock: 153 below, 221 below; Stone: 21 below; Tasker/ICCE: 237 below; Weber: 115 above, 137 above, 203 above, 227 below; Wilmshurst: 55 below; Wisniewski: 83 above, 177 above, 179 above, 185 above; Wothe: 63 above, 69 below, 117 above, 117 below, 119 below, 135 above, 135 below, 145, 171 above, 171 below, 175 below, 209 above, 221 below, spine; Zeininger: 25, 41 above, 77 above, 87 below right, 93 below, 103 above, 105, 115 middle, 219; Ziesler/Angermayer: 83 middle, 83 below, 119 above; Ziesler: 169 above.

INDEX

Acomys minous, 106
Alces alces, 206
Alopex lagopus, 164
Ape, Barbary, 48
Apodemus agrarius, 106
flavicollis, 104
mystacinus, 102
sylvaticus, 102
Arvicola sapidus, 76
terrestris, 76
Atelerix algirus, 10
Aurochs, 212

Badger, 136
Balaena mysticetus, 244
Balaenoptera borealis,
244
musculus, 242
ocutorastrata, 244
physalus, 244
Barbastella
barbastellus, 30
Barbastelle, 30
Bat, Bechstein's, 32
Blasius' Horseshoe, 30
Brandt's, 34
Brown Long-eared,
44
Daubenton's, 34
European
Free-tailed, 46
Geoffroy's, 34
Greater Horseshoe, 28
Greater
Mouse-eared, 32
Grey Long-eared, 44
Leisler's, 40
Lesser Horseshoe, 30
Long fingered, 34
Mehaly's Horseshoe,
30
Mediterranean
Horseshoe, 30
Natterer's 34
Northern, 42
Parti-coloured, 46
Pond, 34
Whiskered, 34
Bear, Brown, 144
Polar, 148
Beaver, European, 80
Canadian, 82
Beluga, 238
Bison, American, 210
European, 210
Bison bison, 210
bonasus, 210
Boar, Wild, 188
Bos primigenius, 212
Buffalo, 210

Canis aureus, 158
lupus, 154
Capra aegagrus, 222
ibex, 218
pyrenaica, 220
Capreolus capreolus,
202
Castor fiber, 80
canadensis, 82
Cat, African Wild, 168
Wild, 168

Cattle, Wild, 212
Cervus axis, 200
elephus, 192
nippon, 198
Chamois, 214
Apennine, 216
Chipmunk, Siberian, 56
Citellus citellus, 62
suslicus, 62
Clethrionomys
glareolus, 68
rufocanus, 68
rutilis, 68
Cotton-tail, Eastern, 118
Cricetus cricetus, 64
migratorius, 64
Crocidura leucodon, 20
russula, 18
suaveolens, 22
Coypu, 84
Cystophora cristata, 180

Dama dama, 196
Deer, American
White-tailed, 198
Axis, 200
Chinese Water, 200
Fallow, 196
Red, 192
Roe, 202
Sika, 198
Delphinapterus leucas,
238
Delphinus delphis, 230
Desmana moschata, 22
Desman, Pyrenean, 22
Russian, 22
Dolphin, Bottle-nosed,
232
Common, 230
Risso's, 236
Rough-toothed, 230
Striped, 230
White-beaked, 230
White-sided, 232
Dog, Raccoon, 166
Dormouse, Common,
92
Edible, 86
Fat, 86
Forest, 90
Garden, 88
Dryomys nitedula, 90

Eliomys quercinus, 88
Elk, 206
Eptesicus nilssonii, 42
serotinus, 42
Equus przewalskii, 186
Erignathus barbatus,
180
Erinaceus algirus, 10
europaeus, 8
concolor, 10
Eubalaena glacialis, 244

Felis silvestris, 168
silvestris lybica, 168
Ferret, 126
Fox, Arctic, 164
Polar, 164
Red, 160

Galemys pyranaicus, 22
Genet, Common, 150
European, 150
Small-spotted, 150
Genetta genetta, 150
Glis glis, 86
Glutton, 134
Globicephala melas, 234
Goat, Wild, 222
Grampus griseus, 236
Gulo gulo, 134

Hamster, Common, 64
Grey, 64
Rumanian, 64
Hare, Brown, 112
Common, 112
European, 112
Irish, 116
Mountain, 116
Tundra, 116
Hedgehog, Algerian, 10
Eastern, 10
Western, 8
Herpestes edwardsi, 152
ichneumon, 152
javanicus, 152
Horse, Przewalski's, 186
Wild, 186
Hylopotes inermis, 200
Hystrix cristata, 110

Ibex, Alpine, 218
Spanish, 220

Jackal, Golden, 158

Lagenorhynchus
acutus, 232
albinostris, 232
Lemming, Norway, 66
Wood, 66
Lemmus lemmus, 66
Lepus europaeus, 112
timidus, 116
Lutra lutra, 138
Lynx, 170
Pardel, 172
Lynx lynx, 170
pardina, 172

Macaca sylvana(us), 48
Macaque, Barbary, 48
Marmot, Alpine, 58
Bobak, 60
Marmota bobak, 60
marmota, 58
Marten, Beech, 132
House, 132
Pine, 130
Stone, 132
Martes martes, 130
foina, 132
zibellina, 130
Megaptera
novaeangliae, 246
Meles meles, 136
Mesocricetus newtoni,
64
Micromys minutus, 100
Microtus agrestis, 72

arvalis, 70
nivalis, 74
Mink, American, 124
European, 124
Mole, 24
Mole-rat, Greater, 62
Lesser, 62
Monachus monachus, 182
Mongoose, Egyptian, 152
Ichneumon, 152
Indian Grey, 152
Small Asian, 152
Monodon monoceros, 238
Mouflon, 224
Mouse, Broad-toothed Field, 102
Harvest, 100
House, 94
Wood, 102
Long-tailed Field, 102
Northern Birch, 108
Rock, 102
Southern Birch, 108
Spiny, 106
Striped Field, 106
Yellow-necked, 104
Muntiacus Reevesi, 200
Muntjac, Chinese, 200
Mus musculus, 94
Muscardinus avellanarius, 92
Muskrat, 78
Mustela erminea, 120
eversmanni, 128
lutreola, 124
nivalis, 122
putorius, 126
putorius furo, 126
vison, 124
Musk Ox, 226
Myocastor coypus, 84
Myopus schisticolor, 66
Myotis bechsteinii, 32
brandtii, 34
cappacicinii, 34
dasycreme, 34
daubentonii, 34
emarginatus, 34
myotis, 32
mystacinus, 34
nattereri; 34

Nonnospalax leucodon, 62
Narwhal, 238
Neomys anomalus, 16
fodiens, 16
Noctule, 40
Nyctalus leisleri, 40
noctula, 40
Nyctereutes procyonoides, 166

Odobenus rosmarus, 184
Odocoileus virginiarus, 198
Ondatra zibethicus, 78
Orcinus orca, 234

Oryctolagus cuniculus, 118
Otter, 138
Ovibos moschatus, 226
Oris orientalis, 224

Phoca vitulina, 174
groenlandica, 180
hispida, 180
Phocoena phocoena, 236
Physeter macrocephalus, 240
Pipistrelle, Common, 36
Kuhl's, 38
Nathusius', 38
Savi's, 38
Pipistrellus kuhlii, 38
nathusii, 38
pipistrellus, 36
savii, 38
Pitymys subterraneus, 74
Plecotus auritus, 44
austriacus, 44
Polecat, Marbled, 128
Steppe, 128
Western, 126
Porcupine, Crested, 110
Porpoise, 236
Procyon lotor, 142
Pteromys volans, 56
Pusa hispida, 180

Rabbit, 118
Eastern Cotton-tail, 118
Raccoon, 142
Raccoon Dog, 166
Rangifer tarandus, 208
Rat, Black, 98
Brown, 96
Common, 96
Norway, 96
Roof, 98
Ship, 98
Rattus norvegicus, 96
rattus, 98
Reindeer, 208
Rhinolophus blasii, 30
euryale, 30
ferrumequinum, 28
hipposideros, 30
mehelyi, 30
Rupicapra rupicapra, 214
pyranaica, 216

Sable, 130
Sciurus carolinensis, 54
vulgaris, 52
Seal, Common, 174
Grey, 176
Harbour, 174
Harp, 180
Hooded, 180
Mediterranean Monk, 180
Ringed, 180
Spotted, 174
Serotine, 42
Shrew, Alpine, 12
Bi-coloured White-toothed, 20

Common, 14
Etruscan, 22
Greater White-toothed, 18
House, 18
Lesser White-toothed, 22
Miller's Water, 16
Pygmy, 12
Pygmy White-toothed, 22
Water, 16
Siscita betulina, 108
subtilis, 108
Sorex alpinus, 12
araneus, 14
minutus, 12
Squirrel, Flying, 56
Grey, 54
Red, 52
Suncus etruscus, 22
Sus scrofa, 188
Souslik, European, 62
Spotted, 62
Spalax microphthal, 62
Stenella caeruleoalba, 230
Steno bredanensis, 230
Stoat, 120
Sylvilagus floridanus, 118

Tadarida teniotis, 46
Talpa europaea, 24
Tamias sibiricus, 56
Thalarctos maritimus, 148
Tursiops truncatus, 232

Ursus arctos, 144

Vespertilio murinus, 46
Vole, Bank, 68
Common, 70
Common Pine, 74
Field, 72
Grey-sided, 68
Northern Water, 76
Ruddy, 68
Short-tailed, 72
Snow, 74
Southern Water, 76
Vormela peregusna, 128
Vulpes vulpes, 160

Walrus, 184
Whale, Blue, 242
Bowhead, 244
Fin, 244
Humpback, 246
Killer, 234
Long-finned Pilot, 234
Minke, 244
Northern Right, 244
Sei, 244
Sperm, 240
White, 238
Wildcat, 168
Wisent, 210
Wolf, 154
Wolverine, 134

Other
Collins Nature
Guides

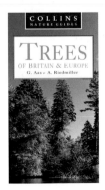

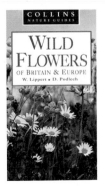

HarperCollins*Publishers*